DATE			

The Nuclear Age

*Power, Proliferation and
the Arms Race*

Congressional Quarterly Inc.
1414 22nd Street N.W.
Washington, D.C. 20037

Congressional Quarterly Inc.

Congressional Quarterly Inc., an editorial research service and publishing company, serves clients in the fields of news, education, business, and government. It combines specific coverage of Congress, government, and politics by Congressional Quarterly with the more general subject range of an affiliated service, Editorial Research Reports.

Congressional Quarterly publishes the *Congressional Quarterly Weekly Report* and a variety of books, including college political science textbooks under the CQ Press imprint and public affairs paperbacks designed as timely reports to keep journalists, scholars, and the public abreast of developing issues and events. CQ also publishes information directories and reference books on the federal government, national elections, and politics, including the *Guide to Congress*, the *Guide to the Supreme Court*, the *Guide to U.S. Elections* and *Politics in America*. The *CQ Almanac*, a compendium of legislation for one session of Congress, is published each year. *Congress and the Nation*, a record of government for a presidential term, is published every four years.

CQ publishes *The Congressional Monitor*, a daily report on current and future activities of congressional committees, and several newsletters including *Congressional Insight*, a weekly analysis of congressional action, and *Campaign Practices Reports*, a semimonthly update on campaign laws.

The online delivery of CQ's Washington Alert Service provides clients with immediate access to Congressional Quarterly's institutional information and expertise.

Library of Congress Cataloging in Publication Data
  Main entry under title:

Nuclear Age

  Bibliography: p.
  Includes index.
    1. Atomic weapons. 2. Atomic energy. 3. Arms race—History—20th century. 4. Antinuclear movement. 5. Atomic weapons and disarmament. I. Congressional Quarterly, inc.
U264.N78 1984   355'.0217   84-5822
ISBN 0-87187-311-7

Author: William Sweet
Editor: Mary L. McNeil
Designer: Mary L. McNeil
Cover: Richard A. Pottern, Patrick Murphy
Graphics: Patrick Murphy
Photo Credits: Cover—Robert Walch/Image Bank; p. 15, National
Portrait Gallery, Smithsonian Institution, Washing-
ton, D.C.; p. 16, Council for a Livable World; p. 71,
Wide World Photos; p. 113, U.S. Department of
Energy; p. 133, 163, U.S. Department of Defense; p.
207, UPI; p. 236, Brad Markel/Capri Press.
Indexer: Toni Gillas

Table of Contents

Tables and Maps

PREFACE

The literature on nuclear issues is massive, but atomic energy, proliferation, and the arms race almost always are treated independently, even though the three areas overlap in many important ways. What we now call the arms race began with the spread of nuclear weapons; and when countries first built atomic bombs they began by constructing reactors, which are essentially the same as the power plants that now produce electricity.

Today the prospects for preventing the spread of atomic arms depend heavily on the outlook for nuclear power. Without understanding something about the technology and economics of reactors, it is impossible to make reasoned judgments about non-proliferation policy. The rate at which countries acquire certain especially sensitive nuclear technologies will have an important bearing, for example, on the very question of whether the spread of atomic arms can be stopped or only slowed.

If proliferation cannot be stopped, nuclear arms negotiations will become increasingly complex and quite possibly hopeless. If, conversely, the nuclear-armed nations remain deadlocked on arms control and determined to keep building their atomic arsenals, they may find it more and more difficult to dissuade non-nuclear countries from joining the club.

Everybody recognizes that there are relationships between strategic arms control, proliferation, and nuclear energy, but too often the connections are treated casually, without adequate attention to detail. The result is that wider ramifications are often downplayed or overlooked entirely. A comprehensive test ban treaty offer is discussed in the context of bilateral strategic arms controls, even though such a measure would have equal or greater importance in the context of proliferation policy. Spent fuel disposal generally is regarded as an environmental problem, even though the fuel contains the crucial materials needed to produce bombs.

I am grateful to Congressional Quarterly for providing me with the opportunity to draw together work on nuclear energy, proliferation, and the arms race, which I have been writing about for newspapers, magazines, and academic publications since the mid-1970s. Mary McNeil, the editor and designer of this volume, was a pleasure to work with. She made many improvements and contributed research for the first chapter. Robert Mudge, a former staff member of Congressional Quarterly, and Robert Benenson provided materials for the section on nuclear waste disposal. Harold A. Feiveson of Princeton University served as consulting editor and gave me indispensable guidance at every stage of the operation. Ann Florini, David Holloway, and David Morrison were helpful with comments and advice on several aspects of the manuscript. Naturally, I alone am responsible for the content of the book.

Acknowledgements customarily end with thanks to the author's spouse, and now that I have written a book I understand why. Without my wife's patience, understanding, and support, I would not have been able to bring this project to a conclusion with my sanity more or less intact.

This book is dedicated to the memory of my uncle, Herbert Grummann, who suffered for several decades from cancer caused by X-ray treatments he was given for a skin condition before World War I.

William Sweet
March 1984

The Nuclear Age

The Nuclear Age

Henry Moore's "Nuclear Energy," erected in 1967, commemorates the 25th anniversary of the first controlled nuclear chain reaction at the University of Chicago.

Chapter 1

THE NUCLEAR ERA IN MID-LIFE CRISIS

The nuclear age is getting to be middle-aged. It has begun to show signs of fatigue, stress, and self-doubt, as though it were a 40-year-old person who cannot decide whether to forge ahead in accustomed paths, turn back, or make some radical new departure.

During the Cold War, when nuclear technologies seemed to be sweeping the world, Albert Einstein remarked that the atomic bomb had changed everything except our way of thinking. If Einstein were to return to life in the mid-1980s, like Rip van Winkle, he might be startled by what he would find. Nuclear technologies have not totally transformed military and industrial practices, contrary to his expectations, but they have begun to have a powerful impact on consciousness and society.

Since World War II some countries have amassed giant arsenals of nuclear weapons, as though another global conflagration rather like World War II were just around the corner. But the arms race has been challenged by a growing international peace movement whose leaders appear to be digging in for a long political struggle. With world opinion continually becoming more hostile to the acquisition of nuclear weapons, countries that once were expected to "go nuclear" at the first opportunity now hesitate to test atomic explosives. Many countries continue to build atomic power plants, which could be used to provide the materials for bombs, but nuclear technology has spread at a much more moderate rate than was anticipated in the early 1970s when it seemed to be the wave of the future.

Faltering Nuclear Plant Construction

In the United States, 80 nuclear power plants were operating at the beginning of 1984, and nearly 50 more were under construction. But utilities had placed no new orders for reactors since 1978 and in just over a decade more than 100 orders had been cancelled. By 1983-1984 nuclear power projects were being abandoned at advanced stages of construction even after billions of dollars had been invested in them. In one instance, which involved the Shoreham plant on Long Island, a bitter political debate erupted over the question of whether it would be worthwhile for a utility to start up a nuclear plant that already was completely finished. Meanwhile, the utility was paying more than $1 million a day in interest on the money it had borrowed to build the facility, which had cost about 10 times more than had been estimated when the project was launched in the early 1970s.

Opponents of nuclear construction have drawn attention to a large number of problematic issues, but of these, perhaps the most damaging to the industry's reputation has been the waste disposal problem. Critics of the industry have considered it scandalous that hundreds of nuclear power plants were built in the United States and abroad before anybody knew how spent fuel from the reactors could be safely stored. In the mid-1970s countries such as West Germany and Sweden enacted laws that barred the construction of additional nuclear power plants until the waste disposal problem was solved. Even in those countries, however, it was not anticipated that permanent underground disposal sites would have been built until well into the 21st century.

Spent reactor fuels contain highly radioactive materials that must be kept completely isolated for up to tens of thousands of years. They also contain plutonium, which can be used as the explosive material in atomic bombs. In the mid-1970s, when the industry assumed that plutonium would be extracted from spent fuels and used as fresh fuel for reactors, experts warned that it would soon be much easier for governments, terrorists, or even criminal groups to obtain material for bombs. If pure plutonium were circulating in large quantities, terrorists or undercover agents working for a foreign government might be able to hijack a shipment and insert the material into pre-fabricated bomb casings.

By the early 1980s it was apparent that plutonium would not be recycled in the United States before the end of the century, but even then the terrorist threat could not be completely discounted. Japan and several European countries have continued to recycle plutonium for commercial reactors. In the five countries that maintain nuclear arsenals, a criminal or terrorist group might conceivably manage to steal plutonium from one of the facilities where it is produced and refined for weapons.

Non-Proliferation Policy at the Crossroads

Many people take the position that it will not be possible to stop the spread of atomic weapons unless all construction of nuclear power plants is terminated and existing reactors are shut down. No doubt nuclear proliferation would be easier to control if everybody agreed to stop relying on reactors to produce electricity. There is little chance, however, that this will happen any time soon.

In spring 1983 more than 200 nuclear power plants were operating in 23 countries outside the United States. Nearly 20 more countries were building commercial reactors or had firm plans to do so. Plans for nuclear construction had been scaled back sharply in the relatively poor nations of Asia, Africa, and Latin America, partly because the types of reactors marketed in the advanced industrial countries have turned out to be much too large to incorporate into small electricity systems. But a number of manufacturers — Kraftwerk Union in West Germany, for example, and France's Technicatome — have launched programs to develop small reactors for export to Third World countries. The developing countries with the most advanced nuclear programs include South Korea, Taiwan, India, Pakistan, South Africa, Argentina, and Brazil, and of these, only South Korea and Taiwan are parties to the Non-Proliferation Treaty (NPT).

The NPT is the linchpin of international efforts to slow or stop the spread of nuclear weapons. Roughly two-thirds of the countries represented in the United Nations have joined the treaty, promising not to develop atomic bombs. Membership in the treaty gives a somewhat misleading impression, however, of the degree to which people have come to rely on nuclear weapons for security.

In addition to the United States, the Soviet Union, England, France, and China, which have assembled nuclear arsenals, U.S. and Soviet allies in Europe and Asia rely ultimately on the "nuclear umbrellas" maintained by the superpowers to guarantee their defenses. When these countries are taken into account, about 2.1 billion people out of a total world population of 4.6 billion can be said to depend on nuclear weapons. Put differently, about 45 percent of the people in the world are in the nuclear-armed camp; if the citizens of India are added to their numbers, the proportion rises to about 60 percent.

In May 1974 India tested an atomic explosive, but it claimed at the time that it intended to use nuclear devices exclusively for peaceful

5

purposes, and during the following decade it did not proceed to mass-produce atomic bombs. The Indian government indicated from time to time that it would permit international inspection of its nuclear facilities if the great powers agreed to far-reaching disarmament measures. But India was not one of the great powers that ordinarily participated in high-level negotiations, and its suggestions often impressed leaders of the nations with nuclear weapons as self-righteous and presumptuous.

Despite India's test, membership in the NPT rose steadily in the 1970s before hitting a plateau in the early 1980s. If the dream of universal membership in the treaty ever is to be realized, however, a bolder diplomacy may be required of the countries that are committed to non-proliferation.

In theory, most of the advanced industrial countries have been unanimous on the importance of preventing proliferation since the early 1970s, but in practice they have found it difficult to stand firm behind common policies. France and Italy, for example, have sold nuclear equipment and materials to countries that supply them with oil, despite qualms about their customers' ultimate intentions.

The deterioration of relations between the United States and the Soviet Union in the late 1970s and early 1980s posed an especially grave threat to solidarity among the countries that support non-proliferation policies. In the 1960s the two superpowers cooperated closely in drafting the NPT, and their joint work on the agreement helped lay the foundation for a period of détente.

An important source of European tension was removed at the end of the 1960s when West Germany ratified a treaty recognizing the borders imposed on the country at the end of World War II. The next major step in the process of East-West accommodation came in 1972 when the superpowers adopted agreements limiting strategic arms. The process of détente reached its high point in 1975 when the two superpowers and 33 European countries adopted the "Helsinki Accords" in Helsinki, Finland. The accords gave formal recognition to the boundaries the Soviet Union had established in East Europe, but they also enunciated the principle that all signatories should honor basic human and political rights.

It was hoped that the Helsinki Accords would open the way for expanding human contacts between East and West, but just a year after their conclusion tensions had begun to mount once again. By the beginning of the 1980s the arms race was back in full swing.

Nuclear War Fighting, Deterrence, and Disarmament

By the end of 1983 the U.S. nuclear arsenal included approximately 26,000 nuclear warheads. About 15,000 of the warheads were deployed on strategic delivery systems (long-range missiles and bombers), and some 10,800 were on short- and medium-range weapons (such as landmines, howitzers, small missiles, and fighter-bombers), or on ships. The short- and medium-range weapons usually were referred to as tactical or theater nuclear weapons because they were intended for combat use in a "theater" of military operations, rather than for strikes against cities or bases in the adversary's home country.

Most of the warheads in the U.S. arsenal as of February 1984 were scheduled for modernization or replacement, and by 1990 the American stockpile was supposed to be increased to include about 29,000 nuclear explosives. According to an estimate prepared by the Center for Defense Information, an organization in Washington that often is critical of Pentagon programs, the U.S. government was planning in 1983 to spend $450 billion on nuclear weapons and support systems over a period of six years.

The Soviets are secretive about their nuclear plans, but their arsenal at the same time was thought to be nearly as large as the U.S. stockpile, and they, too, were modernizing and expanding their strategic and tactical weapons systems. In the mid-1970s they began to deploy a new type of medium-range missile, the SS-20, supplementing older missiles that were targeted at West Europe. NATO responded in December 1979 with a decision to station new U.S. medium-range missiles in five West European countries.

Even before the NATO decision of December 1979, many Europeans were disturbed by the lack of progress in arms control talks and the vast amount of "overkill" in the arsenals maintained by the superpowers. Plainly, each country was maintaining enough nuclear weapons to level the other's cities many times over, and yet both were continuing to refine and increase existing arsenals.

The standard explanation for the size of the stockpiles was that military planners have to be prepared for every kind of contingency, from all-out nuclear war to "limited" exchanges, in which atomic explosives would be used to repulse attacks or to deter escalation of combat. Critics of limited nuclear war doctrines argue that such theories weaken rather than strengthen deterrence, in that they inevitably stir up anxieties that one side or the other might actually launch an attack with the hope of winning. Europeans were particularly alarmed in the first year after President Ronald Reagan took office when administration officials talked openly about the possibility of "prevailing" in nuclear combat. Europeans feared that the superpowers might try to fight out a nuclear war on their soil. Some even believed, or claimed to believe, that Reagan was seriously contemplating a nuclear attack on the Soviet Union.

Among specialists on arms control in the United States and abroad it was widely believed that both superpowers had more than enough nuclear might to inflict unacceptable damage on each other and deter any attack except, possibly, one launched by an utterly deranged leader. The view that any effort to win a nuclear war would be illusory was widely shared by the general public. By the early 1980s concerned citizens were giving more and more support to organizations that were working or claimed to be working to end the arms race. Established peace groups, which had been accustomed to operating on shoestring budgets, found it much easier to raise money, while new groups rapidly gained members, financial support, and attention in the media.

Aware of the growing concern about the arms race, the media did much to highlight the horrors of nuclear war. In 1981 CBS Reports ran a five-part series on U.S. defense policy, which opened with the simulated nuclear destruction of Omaha, Nebraska. On Nov. 20, 1983, ABC aired a two-hour drama, "The Day After," which portrayed the efforts of survivors to cope with the devastating effects of a nuclear attack. The film was followed by a one-hour panel discussion in which the participants disagreed radically on its message and significance.

Carl Sagan, a scientist who recently had participated in a study on the effects of nuclear war, said it was his "unhappy duty to report that the reality is much worse than what has been portrayed in this movie. . . ." Alluding to the conclusions of his study group, Sagan maintained that "even a small nuclear war" would be followed by a "nuclear winter," in which temperatures would be reduced to sub-freezing levels for months. In a "major nuclear war," he said, one or two billion people would be killed directly, and "the biologists who have been studying this think that there is a real possibility of the extinction of the human species from such a war."

William F. Buckley Jr., a conservative writer, commentator, and publisher, said that what viewers had seen was a "hypothetical catastrophe," in contrast to "an on-going catastrophe that is not hypothetical . . .

life in the Soviet Union." Buckley maintained that the "whole point" of the movie was to "launch an enterprise that seeks to debilitate the . . . United States."

Whatever one may think of Buckley's general philosophy, it is indeed hazardous to make hypothetical predictions about the effects of nuclear war. Nobody really knows what kind of effects would result from thousands of interacting atomic explosions. The effort to imagine a catastrophe of such utterly unprecedented proportions boggles even the most sophisticated imaginations. About all that one can say with complete confidence is that any nuclear war would be many times more terrible than what happened at Hiroshima and Nagasaki, where the first atomic bombs were dropped in 1945.

The Birth of the Era

The foundation for the research that led to the development of the atomic bomb was laid in 1905 when Einstein postulated that mass converted into energy and energy into mass in a constant ratio. Einstein's theories revolutionized physics, and during the first decades of the 20th century the world's most brilliant scientists worked feverishly to unravel the mysteries of how mass and energy interact in the atoms that are the building blocks of matter. By the 1930s leading physicists were aware of the possibility that if the atom could be split or "fissioned," vast quantities of energy would be released.

In 1938, on the eve of World War II, two scientists at the University of Berlin — Otto Hahn and Fritz Strassmann — managed to split uranium atoms. Lisa Meitner, a former colleague who had fled to Sweden from the Nazis, concluded correctly when she heard of the experiment that a fission reaction had taken place. Word of the Hahn-Strassmann experiment spread quickly among the leading physicists of Europe, and among the first to grasp its implications was Leo Szilard, a Hungarian physicist who had fled Germany in 1933 when Hitler took power. Fearful that scientists in Nazi Germany would find a way of harnessing atomic energy to build a bomb, Szilard began to mobilize scientists in England and the United States.

In August 1939, one month before the outbreak of World War II, Szilard persuaded Einstein to write a letter to President Franklin D. Roosevelt, warning him of the possibility that "extremely powerful bombs of a new type may be constructed." Einstein's letter, which was dispatched in October 1939, prompted Roosevelt to establish a committee to explore the possibility of an atomic bomb's being built. In December 1941, just before Japan's attack on Pearl Harbor, Roosevelt ordered the establishment of a crash program to develop an atomic bomb — the so-called Manhattan Project. For the next three and a half years scientists from all over the world worked frantically at secret Manhattan Project laboratories, and in July 1945 the first atomic bomb was successfully tested in New Mexico.

On Aug. 6, 1945, at 2:45 a.m. Pacific time, a B-29 bomber called the Enola Gay took off from an island in the Marianas carrying a single bomb known among its makers as "Little Boy" or "Thin Man." Little Boy was 10 feet long, 28 inches wide and weighed 9,000 pounds. For every unit of weight of explosive material that it contained, it was roughly one million times more destructive than anything previously built. Its expected yield was more than seven times greater than all the bombs that the Allies had dropped on Germany in 1942.

At 8:15 a.m. Enola Gay opened its bomb bays over Hiroshima, an industrial city in Japan. Little Boy exploded with a blinding flash about 1,000 feet over the city with a yield equivalent to 12,500 tons of TNT. Immediately, the fireball created by the explosion reached temperatures of several million degrees centigrade, and within a second this fireball had grown to a diameter of 250-300 meters, raising the temperature on the ground under it to roughly 5,000 degrees centigrade. This fireball consumed about one-third of the energy created in the explosion, and the thermal radiation emitted by the ball was sufficiently intense to burn exposed skin at a distance of 2.5 kilometers from the blast center.

The blast also generated a crushing shock wave, which shot outward from the blast center ("ground zero") at roughly the speed of sound. Consuming about half of the bomb's energy, this wave demolished everything in its path up to a distance of about 2 kilometers. A hurricane-force wind followed the shock wave, leaving a vacuum in its wake. Once pressure dropped sufficiently behind the wind, there was a moment of stillness, and then the wind reversed direction and blew back toward the blast center with lesser but still very considerable force.

It was a kind of hell on earth, and those who died instantly were among the more fortunate. Thousands died — vaporized, crushed, or burned. But there were tens of thousands more who were still alive and those who could move began to mill about the city, seeking relief from shock, fire, and pain. Thousands threw themselves into the Ota River, which would be awash with corpses by the end of the day, and hundreds poured into Asano Park, a large private estate that escaped the immediate effects of the bomb. According to John Hersey's memorable reconstruction of the events that day, *Hiroshima*, the people fled into the park "partly because they believed that if the Americans came back, they would bomb only buildings; partly because the foliage seemed a center of coolness and life . . . and partly (according to some who were there), because of an irresistible atavistic urge to hide under leaves."

Initially, of course, people in Hiroshima had no idea of what had happened. Only gradually, as they wandered about the city in search of loved ones, friends, and family, did they realize that the entire city had been destroyed. Only later did they learn that it had been destroyed by a single bomb.

On August 5, 1945, the United States dropped an atomic bomb on Hiroshima, Japan. When this shot was taken, the smoke from the blast had risen 20,000 feet above the city and had spread more than 10,000 feet from the center of the blast at ground level. More than 70,000 people died as a result of the bombing.

Chapter 2

THE WIDENING NUCLEAR DEBATE

Even at the end of the bloodiest war in history, in which atrocities of unparalleled scope had occurred, the news of Hiroshima was shocking and astounding. Before Aug. 6, 1945, the public had no inkling that breakthroughs in nuclear physics were about to revolutionize warfare, and few people had any notion of what the new science and technology involved. About all people could grasp, initially, was that the United States had a bomb that could destroy an entire city.

During the years immediately after World War II, it was difficult to think calmly about the bomb. Americans and their elected representatives tended to hope, contrary to what many atomic scientists told them, that the United States would be able to maintain a monopoly on the weapon for many years to come. Emotions cooled with the passage of time, and as nuclear technology penetrated further and further into society, more and more people felt competent to speak out on atomic policy. But it would be a generation before a broad majority of the public felt comfortable enough with the atom to become seriously involved in nuclear controversies.

Statesmen and Scientists

Until the end of World War II, knowledge of the atomic bomb was confined to top political leaders and scientists working on the Manhattan Project. Within these small circles, it generally was recognized that the new weapon would have revolutionary implications—if it worked. But until the final months of the war, there was little time to think in detail about whether the bomb should be used and how it might affect the postwar world.

The Manhattan Project scientists were absorbed in their challenging work, which they were determined to complete as quickly as possible for fear that scientists in Nazi Germany would beat them to the punch.

13

Political leaders had their hands full with military operations, wartime diplomacy, and economic management. While they had reached some tentative understandings about how to deal with the bomb, no firm decisions had been reached when President Franklin D. Roosevelt unexpectedly died on April 12, 1945.

Harry S Truman, Roosevelt's successor, knew nothing of the bomb when he took office. But during the following months, after Germany was defeated and attention had shifted to the Pacific front, he and his advisers increasingly were concerned about whether to use the bomb against Japan and how the bomb might affect relations with the Soviet Union. At the Manhattan Project, work on the bomb was proceeding at an accelerated pace, and scientists began to discuss the same issues. Among the scientists, a dissenting group developed a point of view that was sharply at variance with the policies of Truman's advisers.

The advisers Truman inherited from Roosevelt tended to take it for granted that the atomic bomb would be used against Japan if the weapon was ready in time to shorten the war. The advisers were much impressed with the leverage the bomb might give the United States in dealing with Russia after the war, and the man who was most impressed of all was James F. Byrnes, a key Roosevelt aide who Truman chose to be his secretary of state after taking office. Some of Truman's advisers worried, however, that a unilateral decision to use the bomb might provoke an arms race with Russia and make it impossible to negotiate a postwar agreement on the international control of atomic energy. Secretary of War Henry L. Stimson, the adviser most worried about provoking an arms race with Russia, persuaded Truman to establish a group of diplomatic, military, and science experts—the so-called "Interim Committee"—to issue recommendations on the use of the bomb.

Most of the Manhattan Project scientists, including J. Robert Oppenheimer, director of the Los Alamos laboratory, tended to favor use of the bomb. But as the war drew to a close, a growing minority questioned whether Japan should be the target of the terrible new weapon that had been developed—they felt—mainly as insurance against a Nazi bomb. The most influential dissenting scientists included Leo Szilard, the Hungarian-born physicist who had persuaded Albert Einstein to write to Roosevelt about the possibility of a bomb in 1939, and James Franck, a German-born physicist who in 1933 had resigned his position at the University of Göttingen in Germany to protest the firing of Jewish professors.

James F. Byrnes, the leader of the southern Democrats in Congress before World War II, was director of war mobilization for Franklin D. Roosevelt. When Roosevelt met with Stalin and other allied leaders at Yalta in early 1945 to discuss the postwar organization of the world, he took Byrnes with him to record the critical conversations. Harry S Truman made Byrnes his secretary of state after taking office, and during the summer of 1945 Byrnes played a key role in decisions involving the atomic bomb.

On May 28, 1945, after failing to be granted a meeting with Truman, Szilard went to Spartanburg, S.C., at the request of the White House to see Byrnes. In this illuminating meeting between two strong-minded individuals who both prided themselves on their frankness, Szilard argued that use of the bomb would provoke an arms race with Russia. Byrnes replied that Russia had no uranium to build a bomb. Byrnes said that if the administration refrained from using the bomb, Congress would wonder why $2 billion had been spent to develop it. Besides, he said, demonstration of the bomb would impress Russia with the United States' might. Szilard was dismayed to find that Byrnes considered it more important to get Soviet troops out of Hungary than to avoid a nuclear arms race.

When Szilard left Byrnes's home—he later said in recollections—he thought how much better off the world might have been if he "had been born in America and became influential in American politics and had Byrnes been born in Hungary and studied physics." [1] Byrnes later wrote that Szilard's "general demeanor and his desire to participate in policy making" left him with an "unfavorable impression." [2]

Three days after seeing Szilard, Byrnes attended a key meeting of Truman's Interim Committee, where he received estimates he had requested on how long it might take Russia to build an atomic bomb. While the Manhattan Project scientists guessed three to four years, the project industrialists thought it would take Russia five to 10 years, and

Leo Szilard was probably the first person in the world to become firmly convinced that it would be possible to build an atomic bomb. Fearing that scientists in Nazi Germany might build the weapon, Szilard persuaded Albert Einstein to write to President Franklin D. Roosevelt in August 1939 about the possibility of the United States building an atomic bomb. Szilard made important contributions to the Manhattan Project, but he opposed the use of the bomb against Japan, and after the war he helped found organizations that since that time have been actively promoting nuclear arms control — the Federation of American Scientists and the Council for a Livable World.

the military manager of the project estimated 10 to 20. Byrnes tended to believe the longer estimates and on June 1 the committee decided to recommend the atomic bombing of Japanese cities. The committee considered but rejected the idea of demonstrating the bomb against a non-civilian target, apparently because they thought a mere test would be unconvincing.

In the following weeks, a committee of dissenting scientists headed by Franck prepared a memorandum opposing use of the bomb on the grounds that it would damage U.S. interests regardless of how relations with Russia developed. When the Franck memorandum failed to sway policy makers, Szilard circulated a petition among the physicists that objected to use of the bomb on moral grounds. But the petition never reached Truman, evidently because the military manager of the Manhattan Project delayed forwarding it to Washington.

The Franck Report

At the end of May 1945, when some of the Manhattan Project scientists were becoming increasingly alarmed about how the atomic bomb might be used, the director of the Chicago laboratory, Arthur H. Compton, decided to set up six committees to discuss and report on the various issues in dispute. The most important of the committees was on "Social and Political Implications," which was chaired by the new émigré German physicist James Franck. In one week the committee prepared a report and submitted it to the secretary of war on June 11, 1945. Excerpts follow.

The scientists on this project do not presume to speak authoritatively on problems of national and international policy. However, we found ourselves, by the force of events, during the last five years, in the position of a small group of citizens cognizant of a grave danger for the safety of this country as well as for the future of all other nations, of which the rest of mankind is unaware.

One possible way to introduce nuclear weapons to the world . . . is to use them without warning on appropriately selected objects in Japan. . . .

Russia, and even allied countries which bear less mistrust of our ways and intentions, as well as neutral countries may be deeply shocked by this step. It may be difficult to persuade the world that a nation which was capable of secretly preparing and suddenly releasing a new weapon, as indiscriminate as the rocket bomb* and a thousand times more destructive, is to be trusted in its proclaimed desire of having such weapons abolished by international agreement.

It must be stressed that if one takes the pessimistic point of view and discounts the possibility of an effective international control over nuclear weapons at the present time, then the advisability of an early use of nuclear bombs against Japan becomes even more doubtful. . . . If an international agreement is not concluded immediately after the first demonstration, this will mean a flying start towards an unlimited armaments race.

Nuclear bombs cannot possibly remain a "secret weapon" at the exclusive disposal of this country for more than a few years.

* A reference to Germany's V-1 rocket, which was inaccurate and rather ineffective.

Throughout the war relations had been strained between the Manhattan scientists and military personnel responsible for security, and the scientists tended to blame the military whenever they felt impeded in their work or cut off from statesmen in Washington. After the war ended with the bombing of Hiroshima and Nagasaki, the two groups soon came into conflict in a legislative dispute over how the development of nuclear power should be administered. On a trip to Washington in the fall of 1945, Szilard learned of proposed legislation—the May-Johnson bill—that apparently would give the military substantial control over atomic energy. Szilard also found on this trip, as he later said in his memoirs, "that for the time being at least the scientists who were regarded as being responsible for the creation of the bomb had the ear of the statesmen." [3] Accordingly, Szilard set about mobilizing the scientists to lobby against the May-Johnson bill.

The scientists succeeded in blocking quick enactment of the legislation, and Congress established a Joint Committee on Atomic Energy under the chairmanship of Rep. James O'Brien McMahon, D-Conn., to consider alternatives. In early 1946 Congress enacted the McMahon Act, which vested control over nuclear research and production in a five-member civilian Atomic Energy Commission (AEC). The bill also provided for creation of a General Advisory Committee (GAC) —a panel of scientists—to provide counsel to the AEC.

The scientists regarded the establishment of the Joint Committee, AEC, and GAC as significant victories, and in the coming years they could count on getting a respectful hearing in the deliberations of these bodies. But the McMahon Act also subjected the scientists to tight secrecy restrictions, reflecting the widespread hope that the bomb was a product of American ingenuity that could be kept an American monopoly for a long time to come. Many of the scientists were not optimistic about the prospects for a long-term U.S. nuclear monopoly, but the secrecy restrictions prevented them from effectively taking their case to the public, and their relations with Truman and his advisers were tenuous.

In 1949 when Russia tested its first atomic bomb, the public's reaction showed deep concern. Truman, as he later said, could not believe that "those Asiatics" had managed to build something so complicated. Before making a public announcement of the Russian bomb, Truman made his science advisers sign sworn affidavits saying that they really believed a successful test had occurred. [4]

Following the Russian test, the administration took up the question of whether to build a hydrogen bomb—a weapon of potentially unlimited power, which administration officials hoped would restore a decisive lead to the United States in the arms race with Russia. The AEC's advisory committee, which was headed by Oppenheimer, opposed crash development of the hydrogen bomb, arguing that a weapon of such awesome destructive force should not be built before arms control possibilities had been explored with the Soviet Union. But Truman overruled the committee and ordered development of a hydrogen bomb in 1951.

Oppenheimer's opposition to the hydrogen bomb probably contributed to his difficulties after Gen. Dwight D. Eisenhower took office as president in 1953. Eisenhower was determined to base national defense even more strongly on the country's nuclear might, and his administration soon made it known that any Soviet attack on Europe would be answered with "massive retaliation."

In 1954 atomic energy officials appointed by Eisenhower brought charges connected with alleged security violations against Oppenheimer. After an extremely controversial AEC proceeding, in which questions were raised about Oppenheimer's character, he was stripped of his security clearances. He could no longer serve as a high-level science adviser to statesmen. Many atomic scientists believed that Oppenheimer was the victim of a political vendetta, and after his dismissal some scientists became more inclined to take their arguments directly to the public. In the mid-1950s, for example, scientists helped generate public pressure for a ban on tests of nuclear weapons.

After Russia launched the first earth satellite in 1957, which prompted a great deal of soul-searching in the United States about whether Americans were falling behind in high technology, especially missile technology, scientists enjoyed a kind of Indian Summer in Washington. Eisenhower relied heavily on them for advice, and when exploratory test ban talks were initiated in the late 1950s, teams of scientists conducted the first rounds of negotiations. But scientists would never again occupy the unique position they held in Washington directly after World War II, when statesmen were inclined to believe that only the physicists who had built the first A-bomb could provide the best advice on the mysteries of atomic policy.

Increasingly during the 1950s and 1960s, scientists were replaced in the halls of power by the people who were involved in designing, building, and selling nuclear technology. Peaceful applications of atomic energy seemed more and more promising to government officials and business executives who had been skeptical in the early years about how quickly engineering problems associated with the new technology could be solved.

Engineers, Entrepreneurs, and Nuclear Salesmen

Private firms had been involved in the Manhattan Project as contractors to build and operate facilities, but after the war companies were not particularly eager to pour money into nuclear research and development. Some scientists and science advisers whose views carried weight at the AEC wondered whether nuclear power plants should be developed at all. Many of them had hoped that the United States and Russia would reach an agreement placing all nuclear facilities under international control. After the two countries rejected each other's proposals at the United Nations in 1946, some scientists concluded that it would be best to restrict the development of atomic energy as much as possible.

Until the early 1950s, the AEC put little emphasis on reactor development, and AEC officials were not optimistic that any nuclear power plants would be built in the near future. Lawrence F. Hafstad, director of the AEC's Division of Reactor Development from 1949 to 1955, emphasized in an article for the April 1951 issue of *Scientific American* the "great variety of possible designs" engineers would have to choose from in developing a viable reactor system. "The problem is not to invent a reactor," he wrote. "The problem is to select one that will yield maximum returns. . . . We have never even designed, much less built and operated, a reactor intended to deliver significant amounts of power economically. Our experience is that in this business estimates cannot be taken at face value and invariably turn out to be too low."

At the time Hafstad described the AEC's program for *Scientific American*, the commission's reactor program included several branches, the most important of which was dedicated to the development of a nuclear power plant suitable for submarine propulsion. The submarine program was of greatest immediate interest to the U.S. government, and it was managed by a talented and hard-driving engineer, naval Captain Hyman G. Rickover. After surveying a large variety of possible designs in the late 1940s, Rickover contracted with Westinghouse and General Electric (GE) to build experimental models. GE worked on a reactor cooled by liquid sodium, and Westinghouse developed a reactor cooled by ordinary water, or "light water" as it is referred to in the nuclear industry. Westinghouse made more rapid progress with its model, and in June 1952 the keel was laid for the *Nautilus*, the world's first nuclear-powered submarine.

By this time the AEC was coming under fire for not pushing commercial nuclear power more forcefully. England had launched vigorous programs to build reactors suitable for production of both nuclear weapons material and electricity, and in 1953 it became the first country to attach an operating reactor to an electricity distribution system. That year, in an effort to recover the initiative in the field of peaceful atomic energy, the AEC arranged for Westinghouse to build— under Rickover's personal direction—a light-water nuclear power plant at Shippingport, Pa., for the Duquesne Power & Light Co. Westinghouse, the manufacturer, and Duquesne, the utility, agreed to finance part of the project.

With international competition on the rise, President Eisenhower promulgated an "Atoms for Peace" program in a speech to the United Nations at the end of 1953. The stated purpose of Atoms for Peace was to promote the use, worldwide, of nuclear technology for electricity production rather than weapons. Eisenhower's assumption was that peaceful and military uses of the atom could be kept separate. In keeping with the new program, Congress rewrote atomic energy legislation in 1954 to facilitate the sale of U.S. reactor technology to foreign countries. In 1955 the AEC announced a "power reactor demonstration program," which provided financial assistance and a number of other incentives to private businesses interested in entering the nuclear business. At the time, some Democrats in Congress criticized the AEC program for not doing enough to encourage U.S. manufacturers. But in fact, the foundation was laid for a successful nuclear sales campaign in both domestic and international markets.*

Rickover's program, as historians later concluded, had helped get manufacturers into the nuclear business, "set new standards of precision and quality in the fabrication of nuclear equipment," and trained "thousands of engineers and technicians," so that there would be "a ready supply of qualified, experienced talent to meet rapidly growing

---

* Years later, Atoms for Peace and related programs would come under a great deal of criticism, from a different perspective, for encouraging countries to acquire the technology needed to build atomic bombs. See Chapter 7, p. 113.

industrial requirements." [5] In 1957 Rickover's Shippingport reactor began to generate electricity, and that year Congress passed legislation removing a major obstacle to wider reliance on nuclear electricity. In the expectation that private insurance companies would be reluctant to insure utilities against nuclear power plant accidents, which might result in thousands of injuries and deaths, Congress adopted a bill that absolved utilities of liability for such damages. The Price-Anderson Act of 1957 established an upper limit of $560 million in compensation for personal damages in the event of a nuclear power plant accident, and it set up a $560 million AEC fund against such an accident. Congress guaranteed up to $500 million in government money to make up the lion's share of the fund.

The nuclear manufacturing industry's big breakthrough came in 1963 when the New Jersey Central Power & Light Co. ordered a full-scale commercial nuclear power plant from General Electric. GE promised to build the Oyster Creek plant for Jersey Central at a contractually guaranteed, fixed price—making this the first reactor to be sold anywhere in the world on an almost strictly commercial basis. The Oyster Creek transaction caused a stir in utility board rooms throughout the United States and Europe, which helped create a boom market for nuclear salesmen. By the end of 1967, U.S. utilities had ordered nuclear power plants adding up to more than 45,000 MW (megawatts) of generating capacity.

The Europeans were reluctant to turn to U.S. nuclear technology, partly because they had invested years of work and vast sums of money in domestic reactor programs, and partly because they resented the dominant position of the United States in high-technology fields in general. But at the end of the 1950s, EURATOM—an organ of the European Economic Community founded to promote nuclear energy— reached an agreement with the United States, which offered the Europeans a number of incentives to build prototype reactors based on U.S. technology.

The U.S. government offered favorable terms, partly because of a hunch that the near-term prospects for nuclear sales were much better in Europe than in the United States. As a panel of the Joint Committee on Atomic Energy put it in 1955,

> The growth of an atomic power program will probably not become significant before 1965. A gap may occur for the power equipment manufacturing industry between present domestic interest

in atomic power reactors and actual sales in substantial volume. If the equipment manufacturers . . . are to be expected to carry forward research and development directed toward making atomic power competitive in the United States, the foreign market for power reactors with its high near-term growth potential may offer a solution to bridging this gap. The potential demand may represent a $30 billion market.[6]

By the early 1960s, utilities in Belgium, France, West Germany, and Italy had decided to take advantage of the EURATOM agreement and build three water-cooled reactors following the Rickover design. After Oyster Creek, private manufacturers and utilities almost everywhere in Europe concluded that "light water" was the wave of the future. They gave up their own programs and began to acquire licenses from U.S. firms to build water-cooled reactors. By the end of the 1960s, the so-called "reactor war" appeared to be ending with a near-total victory for U.S. technology. People connected with the Joint Committee on Atomic Energy, the AEC, the Navy, the manufacturers, and the utilities were inclined to boast of the results they had achieved working in close collaboration with each other. As they saw it, government-industry cooperation plus American know-how was the secret to their success. Little did they know that this partnership soon would be denounced as an "atomic establishment," and that having taken credit for great successes, they would be blamed for failings as well.

Environmental Advocacy and Grass-roots Activism

In the early years of nuclear power, utilities wishing to build atomic power plants occasionally ran into opposition from local groups, but objections rarely had anything to do with the nuclear features of the plants. Citizens worried that plants would mar scenic spots or that hot water and steam from the plants would disrupt river and ocean habitats. While there was some concern about the radiation released from reactors, the levels were too small to seriously alarm most informed people.

The nuclear power controversy as we know it today began around 1970, a time of great social ferment. Largely because of the civil rights movement and the Vietnam War, many Americans became involved in political work, and a spirit of critical self-examination infected the country. It was not long before the new spirit spread to the environment. All over the country—in local communities, state capitals, and in

The Reactor Race: A Chronology

1953   England's Calder Hall gas-graphite reactor is world's first to produce commercial electricity.

1957   Rickover's Shippingport light-water reactor comes into operation.

1959   U.S.-EURATOM agreement to have U.S. firms build demonstration reactors in Europe signed.

1961   German utilities, in first major order for a nuclear power plant, opt for a U.S. light-water model.

1962   Germany's Siemens, a giant manufacturer of electrical equipment, begins work on light-water reactors; but Germany and France discuss joint construction of a gas-graphite power plant at Fessenheim.

1964   Fessenheim project collapses when German utilities opt for more light-water plants.

1965   Last-ditch French effort fails to make the gas-graphite reactor the focus of European Economic Community development program.

1965   Sweden's Asea-Atom receives first order for a boiling-water reactor.

1969   Siemens joins with AEG, another large German electrical company, to form Kraftwerk Union (KWU), which soon becomes West Germany's leading manufacturer of pressurized-water reactors under a Westinghouse license.

1970   Having given up on gas-graphite reactors, French government encourages FRAMATOME, a joint subsidiary of Westinghouse and Empain-Schneider, a steel group, and England's Central Electricity Generating Board (CEG) to develop pressurized- and boiling-water reactors under U.S. license.

1970   Siemens allows Westinghouse licenses to expire.

1973   AEG allows General Electric licenses to expire.

1975   French opt for pressurized-water reactor built by FRAMATOME; Westinghouse's stake is bought out and divided between French government and Empain-Schneider.

1979   England's CEG places its first order for a light-water reactor.

Washington—concerned citizens joined with specialists to form environmental advocacy groups. The typical group, as it emerged in the late 1960s and early 1970s, often included scientists and economists, lawyers and lobbyists, organizers and fund-raisers, writers and public relations experts.

One such group, the Union of Concerned Scientists (UCS), was founded in 1969 by students at the Massachusetts Institute of Technology who were concerned about what they saw as the misuse of technology in Vietnam and in the nuclear arms race. In 1970 a recent graduate of Harvard University named Daniel Ford joined the UCS staff after persuading the group to look into the safety of nuclear power plants. Ford was worried that the standard emergency core cooling system, which was designed for use in the event that the regular cooling system failed, might not work. If the emergency system failed, the reactor core could "melt down" through the foundation of the power plant, resulting in the release of massive amounts of radiation to the environment.

As it happened, when the emergency core cooling system was first tested in a semi-scale model at the AEC's National Reactor Testing Station in November and December 1970, the system did not work as expected. UCS learned of the test results and publicized them the following summer. Meanwhile, UCS challenged AEC safety standards in a number of reactor licensing hearings. In response, the AEC decided to hold national hearings on safety standards in 1972, which gave UCS the opportunity to present its position in a national forum. Presenting the case for UCS, Ford and MIT physicist Henry Kendall argued that AEC standards for the emergency core cooling system should be drastically modified.

Ford and Kendall did not succeed in persuading the AEC to rewrite safety standards, but the hearings turned the cooling system problem into a national issue and contributed ultimately to the reorganization of the AEC. In 1973 the AEC unit responsible for research on reactor safety was separated from the Division of Reactor Development and Technology, and the division's director was fired. These steps did not go far enough to satisfy critics of the AEC, who thought that the agency suffered from a fundamental conflict of interest, inasmuch as it was responsible both for promoting nuclear energy and for ensuring its safeness.

In January 1975 Congress split the AEC into the Nuclear Regulatory Commission (NRC), an independent panel appointed by the president to regulate the nuclear industry, and the Energy Research and Development Administration (ERDA), which would be responsible for advanced reactor development and nuclear weapons production. In 1977 Congress abolished the Joint Committee on Atomic Energy, which critics considered too friendly to the "atomic establishment."

By this time a large number of environmental organizations had involved themselves in the nuclear controversy, and the more groups looked into the issues raised by atomic power, the more they found to worry about. Critical Mass, an anti-nuclear group founded by consumer advocate Ralph Nader in 1974, concentrated largely on the safety issues first publicized by UCS. The economics of nuclear power seemed increasingly unfavorable to staff members at the Council on Economic Priorities (CEP) and Environmental Action (EA), despite claims to the contrary from industry organizations such as the Atomic Industrial Forum and the Edison Electric Institute. The CEP, a liberal economic research outfit in New York, and EA, a Washington-based organization founded by the organizers of the 1970 Earth Day, proclaimed that nuclear power was the "bargain that we cannot afford."

The Natural Resources Defense Council, an elite organization that relies heavily on trained scientists and lawyers, involved itself in rule-making proceedings, reactor licensing disputes, and technical issues connected with the proliferation of nuclear weapons. The spread of weapons also received attention from the Friends of the Earth (FOE), a group that split off from the Sierra Club in 1969 to devote more attention to environmental issues that affect the globe as a whole. The Environmental Policy Center, an offshoot of FOE in Washington, D.C., employed professional staff covering the whole range of nuclear issues: safety and economics, radiation hazards, and—above all—waste disposal.

To many anti-nuclear activists, the question of how to dispose of spent reactor fuel, more than any other single issue, seemed to prove that reactor manufacturers, utilities, and regulators could not be trusted to act responsibly. While safety issues involved highly technical questions, which only experts could argue about with complete confidence, the waste disposal issue seemed clear and simple. Utilities were building and operating dozens of nuclear reactors, even though they had no firm plan for the ultimate disposal of wastes that would remain intensely radioactive and dangerous for thousands of years. Citizens all over the country found this situation scandalous, and by the mid-1970s it was becoming commonplace for people to take rather extreme measures to prevent additional nuclear plants from being built. At a number of sites, demonstrators resorted to acts of civil disobedience, resulting in jail sentences, to dramatize their moral outrage. The most widely reported demonstrations took place in 1976 and 1977 at Seabrook, N.H., a proposed nuclear power plant site on an Atlantic estuary.

Meanwhile, anti-nuclear power activism was spreading on the other side of the Atlantic. By 1977 anti-nuclear demonstrations in Europe were bigger and fiercer than anything seen in the United States. The European campaign began in earnest in 1975 when demonstrators occupied a proposed reactor site in Wyhl, a German village in a wine-making region on the Upper Rhein near France. Local vintners, who worried that steam from the proposed plant would block out sunlight vitally needed in grape production, joined with protesters from all over Germany and France. In 1976 and 1977 tumultuous demonstrations took place at Brokdorf, a proposed reactor site on Germany's Baltic coast. In the summer of 1977, violence broke out in an internationally organized demonstration at a French reactor site near Grenoble, and one demonstrator was killed.

Among people working in the nuclear industry, the passion with which critics attacked reactor sites seemed genuinely baffling. Many industry representatives argued with conviction that demonstrators were confusing nuclear power with nuclear weapons, or using nuclear power as a substitute for nuclear weapons. People did not know what to do about the nuclear arms race, industry representatives suggested, and so they attacked nuclear power plants instead.

To the extent anti-nuclear activists considered this chain of thought, they tended to conclude during the following years that they indeed were concerned about nuclear weapons—but in addition to, not instead of, nuclear power. In 1977-1978, when the U.S. government announced plans to deploy a newly designed "neutron bomb" in Europe, demonstrations against the policy broke out in West Germany and the Netherlands. The neutron bomb seemed singularly offensive to many Europeans because it was designed to kill people while leaving physical structures largely intact.

Initially, the neutron bomb demonstrations brought to mind England's "ban the bomb" demonstrations of 1959-1960 and the protests that had swept West Germany in 1955 when U.S. forces were conducting manuevers designed to train troops in the use of battlefield or "tactical" nuclear weapons. The 1955 protests proved to be a mere flash in the pan, however, while the 1977-1978 demonstrations soon seemed to be the

27

beginnings of a mass anti-nuclear movement. In the summer and fall of 1981, after President Reagan took office and launched a major buildup of U.S. military forces, huge demonstrations against the deployment of additional nuclear weapons in Europe took place in many European cities. Demonstrations in Bonn, Amsterdam, London, Madrid, Milan, and Athens were the largest protest movements since World War II.

Inspired by the Europeans, anti-nuclear activists in the United States organized a similar demonstration in New York, which attracted nearly a million people on June 12, 1982. By then a movement to get the United States and Russia to conclude a bilateral "freeze" on testing, production, and deployment of new nuclear weapons was rapidly gaining ground in the United States. In congressional elections in the fall of 1982, pro-freeze groups worked hard to elect politicians who supported their position, and pro-freeze resolutions passed in nine state referendums. In the spring of 1983, the U.S. House of Representatives adopted a resolution endorsing the principle of a bilateral nuclear weapons freeze.

By the early 1980s, three and a half decades after the end of World War II, the nuclear debate seemed to be coming full circle. In the streets of New York, London, Amsterdam, and Athens, hundreds of thousands of demonstrators were endorsing the position first taken by a handful of scientists in the summer of 1945. They were declaring the nuclear arms race an unacceptable threat to humanity, and they were demanding international action. But it remained to be seen whether they could succeed where Szilard and Franck had failed.

The Changing Nuclear Problem

Ever since scientists first realized that an atomic bomb probably could be built, people always have felt there was a "nuclear problem." But the problem has changed dramatically from decade to decade as different kinds of people have considered it under varying circumstances and from varying perspectives.

To the atomic scientists who built the first bomb, the problem seemed mainly a question of how to prevent a nuclear arms race. But to the politicians who controlled the bomb, it was equally a question of whether to use the bomb against Japan and how to use the weapon to extract diplomatic concessions from Russia. As for the American public, after the first bomb was exploded, its main concern was to prevent other countries from getting the information and technology that had enabled the United States to be the first nuclear-armed power.

In the 1950s, when the public realized that the United States could not maintain a monopoly on nuclear technology, it seemed sensible to at least try to stay ahead of other countries. Insofar as there was a nuclear issue, it was a question of outpacing Russia in the race to build hydrogen bombs and intercontinental missiles and a matter of beating out European competitors in the effort to build a working nuclear power plant.

In the 1970s, when Americans began to take a critical look at the "atomic establishment," the nuclear issue became all kinds of things at once: a question of safety, but also a matter of economics; an issue in practical energy policy, but also a legitimate moral dilemma, insofar as it involved threats to future generations posed by long-lived radioactive wastes and spreading atomic bombs.

As the nuclear debate has widened, it has come to encompass far-reaching questions about the nuclear arms race and nuclear proliferation, power plant safety and public health, energy policy, and economics. The debate remains rooted, however, in the basic equipment used to produce electricity and materials for bombs. To take an intelligent part in the debate, one needs to become familiar with basic facts about how nuclear reactors work.

Notes

1. See S.R. Weart and G.W. Szilard, eds., *Szilard: His Version of the Facts* (Cambridge, Mass.: The MIT Press, 1978), 183-185.
2. See James F. Byrnes, *All in One Lifetime* (New York: Harper and Row, 1958), 284.
3. Weart and Szilard, *Szilard*, 224
4. See Herbert York, *The Advisors: Oppenheimer, Teller and the Superbomb* (San Francisco: W.H. Freeman & Co., 1976), 34. York was a member of the President's Science Advisory Committee for presidents Dwight D. Eisenhower and Lyndon B. Johnson and a member of the General Advisory Committee on Arms Control and Disarmament under presidents John F. Kennedy and Johnson. York attributed the source of this quote to a U.S. senator who was a friend of Truman's.

5. See Richard G. Hewlett and Francis Duncan, *Nuclear Navy: 1946-62* (Chicago: University of Chicago Press, 1974), 382-383.
6. Quoted in Warren H. Donnelly, "Commercial Nuclear Power in Europe," report for the House Committee on Foreign Affairs, 92d Cong., 2d sess., 1972, 31.

Part I

Nuclear Energy

The world's first nuclear reactor was built at the University of Chicago in 1942 by a team headed by Italian physicist Enrico Fermi. The "Chicago Pile" consisted of three basic elements: natural uranium pellets, graphite bricks, and control rods.

Chapter 3

HOW REACTORS WORK

It is sometimes said that a nuclear reactor is just another way of boiling water, and this really is the most straightforward way of thinking about an atomic power plant. In a nuclear plant, as in a coal- or oil-fired plant, water is heated to generate steam, pressure from the steam causes turbines to turn, and the turbines generate electricity. The one big difference between a nuclear plant and a conventional plant is this: In a nuclear plant, controlled atomic reactions involving elements such as uranium or plutonium heat the water, while in a conventional plant an ordinary fire generates the heat.

Most people feel they have an adequate grasp of how one goes about building and sustaining a fire. The procedures may be very simple or quite complex depending on fuels, the scale of the fire, and its purpose. Basically, the procedure is quite similar to what one does when lighting up logs for a fire in the living room or getting ready to broil hamburgers on the back porch. One arranges the fuel in some fashion so that it will come into adequate contact with air, perhaps adding some kindling, newspaper, or lighter fluid so that the fire will catch easier, and then lighting a match to start the fire. One may not be able to explain in scientific terms exactly why the fire catches and keeps burning — why it's better to arrange the fuel one way rather than another, why the lighter fluid catches more easily than the charcoal briquettes, or why the briquette will keep burning once a certain amount of heat is generated — but one knows how to make the fire burn, and to feel reasonably comfortable with the procedures.

To understand a nuclear reactor in a similar way, it is unnecessary to explain scientifically why reactions take place better under some circumstances than others. Unless you are a physicist or an engineer, all you really need to understand is how such reactions take place and how they can be harnessed to boil water.

Nuclear Chain Reactions

All nuclear plants rely on the same basic process, which is called a self-sustaining chain reaction. A chain reaction is easily visualized as being similar to what happens when a billiard ball is hit into a group of balls on a table, causing a series of ricochet collisions. But in a nuclear chain reaction, every time a collision takes place a ball splits into two large pieces and several tiny fragments, some of which hit and split other balls.

Three exceptionally heavy isotopes — uranium 233 (U-233), uranium 235 (U-235) or plutonium 239 (Pu-239) — can be used to fuel chain reactions. In the typical chain reaction, a U-235 atom absorbs a neutron, the neutrally charged part of an atom, and splits into fragments, releasing energy and neutrons; at least one of these free neutrons is absorbed by another U-235 atom, which splits into fragments, releasing still more energy and neutrons; at least one of these neutrons is absorbed by yet another U-235 atom.

On average, 2.5 neutrons are released in every U-235 fission reaction. To create a self-sustaining chain reaction, it must be assured that at least one neutron from each fission reaction is absorbed by another U-235 atom so that a follow-on reaction will take place. In

U-235 Chain Reaction

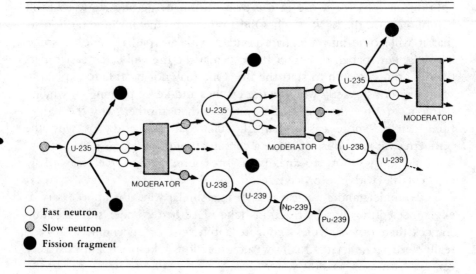

- ○ Fast neutron
- ◐ Slow neutron
- ● Fission fragment

theory, this is a straightforward engineering problem, but a number of factors make it somewhat difficult to solve.

The neutrons released in U-235 fission reactions are highly energetic "fast" neutrons, which unless they are quickly slowed to lower energies will tend to be captured by U-238, a heavier isotope, before they have a chance to initiate another fission. Also, some of the neutrons will tend to escape from the reactor vessel. As it happens, natural uranium consists almost entirely of the 238 isotope,* and only 0.7 percent is 235. Because the proportion of 238 is so high, neutrons released in natural uranium fission reactions stand little chance of being captured by other 235 atoms, and that is why self-sustaining chain reactions do not take place in natural deposits of uranium.

To generate a self-sustaining chain reaction in uranium, engineers can take one of two paths or some combination of the two. They can either find means of increasing the proportion of the 235 isotope in the fuel so that neutrons stand a better chance of being captured by the right atoms; or they can find a means of "moderating" the energy of the fast neutrons so that the neutrons will be less readily absorbed. A good moderator, such as deuterium (heavy hydrogen) or graphite, minimizes the probability that neutrons will be absorbed by the U-238 and maximizes the probability that they will be absorbed by U-235.**

If a reactor is designed to operate on natural uranium, an especially good moderator is needed to convert fast neutrons into "thermal" neutrons. If, on the other hand, a special fuel is prepared that consists entirely, or almost entirely, of U-235 or some other fissionable material, a moderator can be dispensed with entirely. Reactors of both types have been built and continue to be built, but most commercial power plants incorporate aspects of both designs.

The world's first nuclear reactor, which was built by Enrico Fermi's team at the University of Chicago in 1942, was a natural uranium device moderated with graphite. The nuclear power plants of the future may be

---

* Isotopes are forms of the same element that have almost identical chemical properties but different atomic weights. Isotopes of an element have the same atomic number, which is equivalent to the number of (positively charged) protons in their nuclei, but the number of neutrons in their nuclei varies.

** Hydrogen has the atomic number 1 and an atomic weight of 1.00797. Deuterium, a hydrogen isotope, has the atomic number 1 and an atomic weight of 2.0141. Both ordinary hydrogen and deuterium are effective moderators. But because deuterium already has a neutron, it is far less likely than ordinary hydrogen (or water) to absorb neutrons itself.

"fast reactors" or "breeders," which do not need a moderator. But the reactors that produce commercial electricity today all rely on moderators, and they are fueled with natural uranium or uranium in which the concentration of U-235 is somewhat elevated.

Commercial Reactor Types

The reactors that currently produce electricity, whether they were built in 1960 or 1980, easily are recognized as the same kind of device, just as an automobile engine built in 1920 readily is seen to be the same type of machine that one finds in a car coming off the assembly line today. Every reactor is arranged in roughly the same way, and every reactor contains the same key components.

The *fuel* used in all the reactors that have become commercially viable since World War II is uranium, either in its pure metallic form or in the form of uranium oxide (UO_2). Generally, the fuel pellets are inserted

Pressurized-Water Reactor

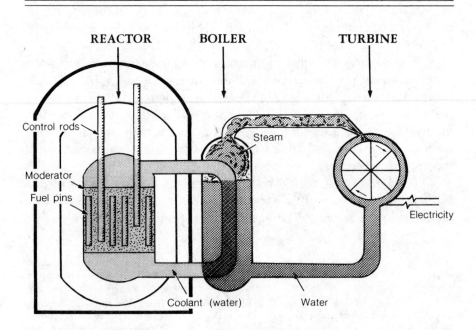

REACTOR   BOILER   TURBINE

Control rods

Steam

Moderator

Fuel pins

Electricity

Coolant (water)   Water

into metal tubes, which are arranged in a lattice to form the reactor core. The *moderator,* which can be graphite or some other form of carbon, ordinary water, or heavy water,* is distributed evenly throughout the reactor core so that neutrons emanating from fuel rods stand a good chance of being slowed down before they reach other fuel rods. *Control rods*, made of some substance such as cadmium that has a strong tendency to absorb neutrons, can be inserted into the reactor core to regulate the reaction.

A *coolant,* which doubles as moderator in some reactors, is piped through the core to remove heat to the turbine system. The coolant powers the turbines directly in certain reactor types, while in others, *heat exchangers* transfer energy to a secondary system that drives the turbines. Backing up the normal cooling systems, for use in the event there is a rupture in the pipes or some kind of control error, is an *emergency core cooling system.* Surrounding the core, the piping system, and sometimes the turbines as well, is a more or less elaborate *containment structure* designed to prevent leakage of radiation during normal operations and to reduce radiation releases in the case of a major accident.

The first large-scale reactors were built during World War II to produce plutonium for atomic bombs.** They essentially were the same as the Chicago reactor in that they were composed of graphite piles laced with uranium slugs, but they were much larger — about five stories high. Because they generated much more heat than the small Chicago pile, they needed to be cooled. The designers of the reactors — Manhattan Project scientists and engineers — originally intended to cool them with helium gas, but when this proved impractical they relied on fresh water from a nearby stream. The reactors were built in a remote part of Washington state at a specially prepared site, the Hanford Nuclear Reservation, which has remained a major U.S. military installation. Updated versions of the original graphite piles continue to produce materials for U.S. nuclear weapons at Hanford, and most reactor wastes from the military program are stored there.

The first reactors built overseas after World War II closely resembled the Hanford piles, except that turbines soon were added to generate electricity on an experimental basis. Among the allied countries

---

* In heavy water some of the molecules consist of oxygen and deuterium — the heavier hydrogen isotope — instead of regular hydrogen. The molecular formula for water is of course H_2O); for heavy water it is D_2O).
** When U-238 atoms absorb neutrons from U-233 fission reactions, it is converted into U-239, which decays into Pu-239.

that had defeated Germany, the main priority was to acquire the technology needed to build atomic bombs. But there also was interest in exploring the possibility that reactors might be suitable for energy production.

Russia is thought to have tested its first small-scale power reactor in 1954, the same year Rickover's team started up the prototype reactor for the first U.S. atomic submarine. Russia's first full-scale power plant began operating in 1958, just a year after Rickover's Shippingport reactor began operation. Like the Hanford reactors, the first Soviet power reactors were graphite piles cooled by water.

The British and French governments, unlike Russia's, remained on friendly terms with the United States after World War II. But the British and French did not entirely trust the United States, which had retreated into isolationism after World War I. And, because their scientists had contributed to the success of the Manhattan Project, they bitterly resented being cut off from access to U.S. nuclear information after World War II. The British government, eager to remain a major power in world affairs, resolved to build an independent arsenal of atomic bombs soon after the war. In France, a succession of weak governments deferred the question of whether to build bombs, but they saw to it that the relevant technology was assembled so that bombs could be built quickly if and when a weapons program was considered advisable.

The first British and French reactors were graphite piles fueled with natural uranium and cooled with carbon dioxide. Like the Russian reactors and the Hanford piles, their purpose was to convert U-238 into Pu-239 for use in the manufacture of atomic bombs.

In the United States the Manhattan Project had developed the technology to produce bombs both from Pu-239 and from U-235, but the method employing plutonium had proved to be easier and cheaper. It turned out to be extremely hard to "enrich" uranium — to boost the U-235 isotope relative to U-238 — and the enrichment facility built at Oak Ridge, Tenn., was in fact the largest and most expensive industrial facility ever built.

Still, the existence of the Oak Ridge enrichment complex opened possibilities for U.S. reactor designers after World War II that were not available, initially, to their counterparts overseas. For example, a reactor could be built more compactly if it was to be fueled by enriched uranium rather than natural uranium. Ordinary water, a relatively weak moderator, could be used rather than graphite or heavy water. These were decisive advantages in the eyes of Rickover, as he contemplated alternative designs for the first U.S. submarine reactor.

The reactor developed by Rickover's team, the so-called "light-water reactor," relied on ordinary water both to cool the core and to moderate the reactions fueled by slightly enriched uranium. Rickover's team thought it would be disadvantageous to allow the water to boil, so they designed the reactor to keep the water under pressure, which reduces its boiling point.

In the version of the "pressurized water reactor" built by Westinghouse under Rickover's direction at Shippingport, Pa., primary and secondary cooling systems are connected by heat exchangers so that the turbines never come into direct contact with water that has been irradiated. The Shippingport model, which has proved to be the prototype for all the reactors Westinghouse has sold since 1957, is fueled by "low-enriched uranium," that is, uranium in which the proportion of the U-235 isotope is boosted from less than 1 percent to just more than 3 percent.

The variant of the light-water reactor developed by General Electric (GE) during the late 1950s and early 1960s also is fueled by low-enriched uranium, but the GE reactor permits the cooling water to boil and drive the turbines directly. In some respects, GE's "boiling water reactor" is simpler and cheaper than Westinghouse's pressurized-water reactor because it does without heat exchangers and a secondary cooling system. But the reactor vessel is much larger in the GE plant, refueling is a more cumbersome process, and the containment building has to enclose much more equipment because the turbines come into direct contact with water that has passed through the radioactive core. Overall, the pressurized-water reactor and boiling-water reactor have turned out to be about equally expensive to build and operate.

When the pressurized-water and boiling-water reactors first appeared on the market in the late 1950s, it was not obvious that utilities would prefer them to models being developed in other countries. In contrast to other reactors, they depended on enriched uranium, an expensive fuel available — initially — only from the United States. To be refueled, they had to be shut down for a period of time. And because they operated at relatively low temperatures, their thermal efficiency — the efficiency with which they convert energy into electricity — was rather low.

In England, engineers, manufacturers, and energy officials remained confident well into the 1960s that their reactor lines would prevail in the end. They thought their first commercial model, the MAGNOX reactor, performed satisfactorily. It was a natural uranium-fueled, graphite-moderated, and carbon dioxide-cooled reactor; the name "MAGNOX" was derived from a special neutron-resistant magnesium alloy that was invented to clad the fuel.

The second commercial model under design in England, the advanced gas reactor, was meant to achieve improved efficiencies by operating at higher temperatures. It used uranium oxide rather than uranium metal as fuel and stainless steel rather than the magnox alloy for the fuel cladding because uranium metal and magnox would melt at the reactor's operating temperatures.

Engineers in England, West Germany, and the United States hoped to achieve still higher efficiencies by building a "high-temperature, gas-cooled reactor," which would dispense with metals in the reactor core and rely on ceramic cladding materials instead. The high-temperature reactors developed in the 1960s used highly enriched uranium as fuel, or uranium in which the proportion of U-235 is increased to more than 90 percent. Designs for high temperature reactors anticipated, in the late 1960s and early 1970s, that thorium also could be used as a fuel. Thorium 232 (Th-232) absorbs neutrons to become thorium 233 (Th-233), a fissionable material, much as U-238 absorbs neutrons to become Pu-239. If combined with U-235 in the reactor core, Th-232 would be converted into U-233, which in turn would fission, contributing to the chain reaction. Because thorium is quite plentiful, its potential use in the high-temperature reactor was considered a major advantage.

Prototype high-temperature reactors were built in England and the United States, and West German engineers developed a design that departed radically from all previous reactor models. In the so-called "pebble bed" reactor they hoped to build, the fuel would be molded into balls about the size of billiard balls, which would be coated with a moderator, graphite. The balls would be fed from the top into the reactor vessel, much as coke is fed into a blast furnace, and as the fuel was exhausted, the balls would be removed from the bottom. Helium gas, which transfers heat more efficiently than water, would be blown up through the balls from the bottom to cool the reactor and transfer heat to a turbine system or to industrial uses.

Hopes for the high-temperature reactor largely were dashed in late 1975 when General Atomic, the company that had committed itself the most to the development of such reactors, announced that it was terminating all work on the project. The withdrawal of General Atomics, a subsidiary of Gulf Oil and Royal Dutch Shell, was a trouble signal for the companies and research teams that were at work on advanced reactor designs. By this time, utility managers in Europe as well as the United States were showing an overwhelming preference for the light-water reactor. European manufacturers, who acquired light-water reactor technology under license from Westinghouse and GE, already were preparing to operate completely independently of their American sponsors. Two other U.S. companies — Babcock & Wilcox and Combustion Engineering — also had begun to build light-water reactors.

Even in Russia, engineers had adopted a design using water as the coolant-moderator and low-enriched uranium as fuel. The Russian design was so similar to Westinghouse's pressurized water reactor that it came to be called the "Eastinghouse" reactor.

By the mid-1970s, only Canada's natural-uranium, heavy-water reactor was surviving in the world market as a viable competitor to the light-water reactor. Canada's scientists and engineers had specialized in procedures using heavy water in the Manhattan Project, and after the war they set to work on a reactor that would be moderated by D_2O. The first full-scale model of this reactor came into operation in 1971.

Dubbed the "CANDU," for "Canada-deuterium-uranium," the heavy-water reactor can be refueled continuously without being taken out of operation, unlike light-water reactors. The CANDU runs on natural uranium, a significant advantage for Canada, which has abundant uranium reserves and the equipment to produce heavy water but no uranium enrichment facilities. For most other countries, however, this advantage is less decisive because heavy water is difficult and expensive to produce, although it is not as hard to make as enriched uranium.

Fast-Breeder Reactors

When reactors first were conceived and designed, scientists and engineers widely assumed that nuclear power could not be a viable energy source over the long haul unless "fast reactors" or breeders were developed. Because natural uranium consists largely of non-fissionable U-238, which is wasted in water-moderated plants, it seemed that some

Fast-Breeder Reactor

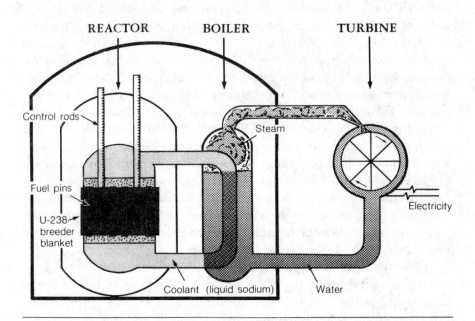

way had to be found of utilizing the 238 isotope if nuclear power was to be a long-term replacement for oil and coal. Breeder reactors seemed to be the answer.

In a breeder the basic idea is to surround a core of fissionable U-235, Pu-239, or U-233 with a "blanket" of non-fissionable U-238 or Th-232. As fast reactions take place in the core, neutrons are ejected into the blanket where they are absorbed by the "fertile material." The neutrons convert the U-238 or Th-232 in the blanket into fissionable Pu-239 or U-233, which can be extracted and used as new fuel for the core.

The hope is to develop breeders that will produce more new fuel from U-238 or Th-232 than they consume in the form of U-233, U-235, or Pu-239. In theory, breeders operating on the uranium-plutonium cycle could stretch the amount of energy yielded by a given amount of uranium by a factor of 30-100. Breeders operating on the uranium-thorium cycle could stretch reserves of radioactive materials even further and give nuclear energy a still longer lease on life.

In the late 1960s and early 1970s, the effort to develop a commercial breeder began to resemble an international competition, in which victory was expected to bring substantial economic advantages. The winning country would stand a chance of dominating the world market for second-generation reactors, and moreover, the winner's high-technology industries could be expected to benefit from association with a breakthrough in one of the most advanced engineering fields. In the eyes of some contestants in the breeder race, the glory of winning began to seem almost more important than the prospect of immediate material gain.

By the mid-1970s, a diverse group of countries, including India, Italy, Japan, and West Germany, had launched breeder programs. But the most advanced programs were in England, France, and the Soviet Union — precisely the countries that lost out to the United States in the race to dominate the market for the world's first generation of nuclear power plants. Of these three countries, France generally was recognized by the end of the 1970s as having the world's most advanced breeder program.

The French government ordered construction of the first full-scale commercial breeder reactor in late 1977. Operating experience with two earlier breeders, the 40 MW Rapsodie and the 250 MW Phenix, was deemed satisfactory, and French engineers expressed confidence that they were ready to go ahead with a 1,200 MW "Superphénix."

The Superphénix, like all the latest breeder models, is a sodium-cooled fast reactor. Liquid sodium, or some similar substance that can absorb and transfer heat extremely efficiently, has to be used to cool breeders because of the very high temperatures generated by the core reactions. The Superphénix differs from some other recent fast reactor models in that it follows the "pool" rather than the "loop" design. In the pool design, the reactor core, primary sodium pumps and heat exchangers, are all contained inside one large vessel, while the pumps and heat exchangers are outside the vessels in "loop" models. The loop design facilitates operation and maintenance, but the French found the pool to be slightly superior than the loop design when they adopted it for the Phénix.

The fuel pins in the Superphénix are similar to the hexagonal pins designed for the Phenix. Fuel pins located around the perimeter of the core consist entirely of fertile material, while the central pins contain fissionable material in the middle — a mixture of U-235 and Pu-239 — and fertile material at either end. Thus the whole bundle of fuel assemblies approximates the shape of a sphere, with a core of fissionable material surrounded in every direction by fertile material.

Writing in *Scientific American* in March 1977, the director of the French Superphenix project boasted that "not a single pin [had] failed while in service in [the] Phenix," and he claimed that "the cost of the kilowatt-hour of electricity produced by [the] Superphenix will be in the same range as that produced by an oil-fired station." He conceded, however, that small sodium leaks detected during the summer of 1976 required the Phenix to be shut down for repairs. In April 1982 the Phenix sprang a leak in its cooling system, and when sodium came into contact with air, a fire broke out. A similar accident in a Russian intermediate-scale breeder reactor is thought to have caused a major explosion in 1974.

The use of sodium, which explodes spontaneously in contact with air, is just one of the problematic features connected with contemporary breeder designs. By nature, breeders are hard to control because reactions take place at a very high rate, and any flaw in the cooling or control system could lead to catastrophic results. In contrast to reactors that operate on natural or low-enriched uranium, which have an innate tendency to stop reacting if the core is damaged, breeder reactors can speed up in accidents, causing — in the worst conceivable case —an atomic blast.

Breeders consistently have proved to be more difficult to build and operate than engineers had hoped. The United Kingdom's 250 MW breeder reactor at Dounreay, Scotland, is reported to have been in operation less than 10 percent of the time during its first eight years in operation. A 300 MW West German breeder, which Belgium, the Netherlands, and the United Kingdom are helping to finance, is thought to be about eight years behind schedule, and the estimated costs of the project have increased fourfold. Even the Superphénix is expected to cost three times as much as its builders expected five years ago, and it will be at least twice as expensive as a conventional nuclear reactor of comparable output. The French government has announced it may build only one more breeder reactor after the Superphénix, instead of the five originally planned, and top French energy officials now concede that fast reactors may not be commercially attractive until after the year 2000.

When the French launched the Superphénix project, it was widely assumed that construction of conventional nuclear power plants would steadily accelerate as countries sought to reduce their dependence on oil. Uranium prices would rise sharply, the reasoning went, and the economics of alternative nuclear fuels would look increasingly attractive. What in fact has happened, however, is that nuclear construction has lagged far behind projections, and uranium prices have stabilized or dropped. As a result, there is no urgent need to replace natural uranium with Pu-239 or U-233 produced in breeder blankets, and there will be little need before the end of the century.

In the United States a project to build a liquid-sodium breeder reactor at Clinch River, Tenn., has been mired in controversy since the mid-1970s. Initially, citizens concerned about public safety and environmental impacts held the project up. After President Jimmy Carter took office, he tried to terminate the project, partly because it was beginning to look like an unattractive investment, but mainly because he sought to discourage use of plutonium as reactor fuel. Carter and his advisers reasoned that if use of plutonium became widespread worldwide, it would become extremely difficult to prevent other countries or terrorists from using it to build atomic bombs.

The dangers Carter attributed to plutonium seemed far-fetched to supporters of Clinch River, who felt it would be a grave error for the United States to allow countries like France and the Soviet Union to dominate advanced breeder technology. Many members of Congress, especially members of the Tennessee delegation, were loathe to abandon a project in which hundreds of millions of dollars already had been sunk. Congress kept Clinch River just barely alive during Carter's term, and Ronald Reagan took office with the intention of resuscitating the project. Still Congress remained closely divided on the issue until October 1983 when the Senate voted for the first time to deny any further funding for Clinch River.

'Once-through' and Plutonium Fuel Cycles

The term plutonium economy, which frequently is used pejoratively, refers to a nuclear energy system in which plutonium is extracted from spent reactor fuels and recycled for use as fresh reactor fuel. The equipment used to extract plutonium from spent reactor fuels is called "reprocessing" technology. Reprocessing is considered a highly sensitive technology, like enrichment, because it can be used to get fissionable materials needed for construction of atomic bombs.

In principle, plutonium can be recycled for use in conventional atomic power plants regardless of whether breeders are built, and until

45

recently it was assumed in the nuclear industry that light-water reactor fuels would be recycled. Recycling would reduce uranium requirements, and it might also help with waste disposal — or so it was thought. However, it would cost more to extract plutonium from spent fuel using the current U.S. reprocessing technology than the substance is worth as fresh fuel. Moreover, it has become apparent that reprocessing is likely to complicate final waste disposal about as much as it helps it.*

Reprocessing involves separating extremely radioactive materials, which are dangerous to human health, highly corrosive, hot, and explosive. The materials have to be handled exclusively by remote control, in heavily shielded chambers, much of the time under water. While reprocessing is done as a matter of course in the nuclear weapons countries to produce plutonium for atomic bombs, all efforts to make reprocessing technology work on a commercial basis have run into serious difficulties.

The first U.S. plants, which were located in upstate New York and Illinois, had to be closed after engineering problems could not be solved satisfactorily. A much more advanced reprocessing plant was built at Barnwell, S.C., in the early 1970s, but Carter withdrew government support from the project and prohibited the plant from being opened. Reagan's first secretary of energy, a former governor of South Carolina, sought renewed congressional support for the project, but without success.

England and especially France, despite serious difficulties with their reprocessing facilities, are determined to proceed with commercialization of the technology. Together with a consortium of West German chemical companies, the French and British governments have established an international holding company called United Reprocessors GmbH (URG), which has begun to handle spent fuels. A number of countries including Switzerland and Japan have sent spent fuels to the URG facilities at La Hague, France, and Windscale, England for reprocessing. U.S. officials and members of Congress who are concerned about the spread of plutonium have agonized about how far to go with efforts to inhibit URG operations.

Even in Europe and Japan, it is unlikely that plutonium will be recycled on a large scale unless and until breeder reactors begin to be

---

* See Chapter 5.

Nuclear Fuel Cycle (Once-Through)

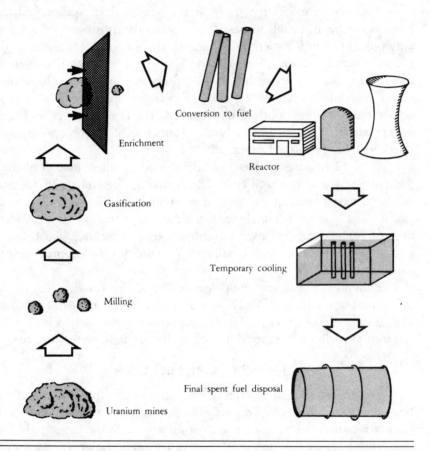

Conversion to fuel

Enrichment

Reactor

Gasification

Temporary cooling

Milling

Final spent fuel disposal

Uranium mines

deployed in significant numbers, which will not happen before the end of the century at the earliest.

The most common once-through cycle includes the following steps: mining of uranium; milling of uranium and manufacture of "yellowcake"; uranium enrichment; fuel fabrication; irradiation of fuels in reactors; cooling of spent fuel; and processing of spent fuel for final disposal. The steps in this cycle that can cause the most serious difficulties are final disposal, which has yet to be worked out, and enrichment, which only a few countries are equipped to do. The United States, for its part, is sure to stick with the "once-through" fuel cycle at least until the 1990s.

The five countries that have arsenals of nuclear weapons all have built enrichment facilities based on "gaseous diffusion" technology. Gaseous diffusion is a method of filtering uranium in giant plants so that U-235 becomes more concentrated. Urenco, a German-Dutch-British consortium, has developed a more advanced enrichment technology based on centrifuges. A new U.S. enrichment plant, which is being constructed at Portsmouth, Ohio, employs the centrifuge technology. It is expected that a still more advanced enrichment technology involving lasers will be developed eventually.

West German engineers have developed an alternative enrichment technology that is not well described in public literature, and Germany has made controversial sales of the technology to South Africa and Brazil. The German enrichment technology generally is considered inferior for commercial use, and buyers of the technology are widely suspected of wanting it primarily to provide for an atomic weapons capability.

Enrichment facilities are not needed to service reactors that operate on natural uranium — CANDUs, for example, or the early French and British gas-graphite reactors. Otherwise, however, the fuel cycles for such reactors virtually are identical to the cycle for light-water reactors.

Importance of Fuel-Cycle

For engineers, manufacturers, and utility executives, it naturally has been important to think carefully about what types of reactors to build once it was decided to launch nuclear programs. In the larger public debate about nuclear power and nuclear weapons, however, issues connected with the fuel cycles are more important. The question of whether to proceed with commercial plutonium recycling is central to current nuclear proliferation policy, and curbing the spread of atomic bombs depends on close attention to the points in the fuel cycle where enriched uranium or plutonium can be diverted for use in weapons. As for nuclear power, its destiny ultimately may depend on the economics of plutonium recycling and the question of whether spent fuels can be disposed of safely and at reasonable expense.

Commonwealth Edison's twin-unit LaSalle County nuclear power station near Marseilles, Ill. Despite the utility's reputation for producing cheap nuclear electricity, it was dealt a severe blow in 1984 when the Nuclear Regulatory Commission refused to allow it to open a similar plant because of quality-control problems.

Chapter 4

NUCLEAR ECONOMICS

In 1973 when the Organization of Petroleum Exporting Countries (OPEC) boosted the price of oil fourfold, it generally was assumed that industrialized and industrializing regions of the world would switch wholesale to nuclear power. The OPEC price hike made it decisively cheaper to rely on uranium rather than oil for electricity, except in countries that had their own petroleum reserves. Coal was in short supply in many parts of the world, and besides, coal seemed old-fashioned and dirty. Whatever was said about the health and safety hazards associated with nuclear energy, the risks connected with mining and burning coal seemed hardly better.

World Reaction to Oil Crisis

In the years immediately following the 1973 oil embargo, trends in many countries seemed to bear out the view that nuclear energy represented the wave of the future. Brazil, a fast-growing country with a Gross National Product (GNP) that ranks among the world's top 15, placed a multibillion-dollar order with West German manufacturers for reactors and equipment to make and process fuel for nuclear reactors. When India, another of the top 15 countries, tested an atomic bomb in 1974, attention was drawn to the country's ambitious nuclear plans — a program that involved development of breeder reactors and facilities to service a plutonium fuel cycle, as well as conventional light-water and heavy-water reactors. During the summer following India's test, Egypt and Israel tentatively agreed to place large nuclear orders with U.S. firms, while countries such as Iran, South Korea, Taiwan, and Argentina stepped up their nuclear construction programs.

In Western Europe, governments almost universally reacted to the OPEC "oil shock" by announcing accelerated nuclear programs. Because indigenous oil reserves virtually are nonexistent in Europe, and coal

51

reserves are substantially depleted, officials reasoned that there would be no serious alternative to "going nuclear." Energy conservation, a good substitute for added energy use in countries such as the United States, seemed much less promising in most of Western Europe. The Europeans for a long time had been much more careful than people in the United States to consume energy efficiently and sparingly, and when the OPEC oil shock hit them in 1973-1974, their per capita consumption of energy was, on average, about half the U.S. level.

By the late 1970s it was becoming apparent that the switch to nuclear power would be more difficult than European energy officials initially imagined. In West Germany, reactors proved to be as controversial they were in the United States, and by the beginning of the 1980s construction of new nuclear power plants had come to a virtual standstill. In England, exceptionally harsh economic problems crippled nuclear construction. At the same time energy officials became mired in an argument about whether to stick with native technology or switch over to light-water reactors designed in the United States. Demand for new electrical generating plants lagged far behind expectations everywhere in Europe — largely because of a stubborn economic recession induced by high oil prices.

Even so, governments forged ahead with nuclear construction programs in a number of European countries — sometimes amidst heated public controversy, sometimes almost without open debate. By the end of 1982 Finland — the world's leader in nuclear construction — was relying on atomic power plants for 40 percent of its electricity. France was producing 38.7 percent of its electricity from nuclear power, and the government expected atomic plants to supply 70 percent of the country's electricity by 1990. Sweden produced 38.6 percent of its electricity from nuclear power at the end of 1982.

In Sweden, all aspects of the nuclear issue were discussed systematically at the community level during the late 1970s; the socialists, who had been in power for 44 years, were defeated in 1976 largely because of their support for nuclear energy. Finally, in 1980 the question of whether to proceed with nuclear construction was submitted to a national referendum, which produced inconclusive results. Nearly 39 percent of the Swedes voted to terminate nuclear construction and to phase out existing reactors within a decade; just more than 39 percent voted to complete six new reactors, but to phase out nuclear power totally during the next 25 years. In France, which has had a strong tradition of central planning since the time of Louis XIV (1638-1715), energy officials have responded little to sporadic anti-nuclear protests.

U.S. Position: Nuclear vs. Coal

Nuclear energy is only used to produce electricity. For most uses, it is not practical or economical to replace oil and gas with electricity as an energy source. For example, automobiles cannot be run using electrical energy. Nuclear energy, therefore, can replace oil only to the extent that oil is used to produce electricity, as a general rule, and in the United States oil has not been used very much in electric power plants.* Even before the energy crunch, coal was the most important single source of electricity in the United States.

In 1973 coal-fired plants accounted for 45.6 percent of all U.S. electricity production, and oil for 16.9 percent. By 1981 oil had dropped to 9.0 percent, while coal had climbed to 52.4 percent. Nuclear-generated electricity increased from 4.5 percent in 1973 to 11.9 percent in 1981. Even then, however, nuclear plants met less than 4 percent of total U.S. energy needs, while oil accounted for 44 percent of the nation's energy consumption — hardly different from 1973 when it accounted for 46.7 percent.

Given the country's bountiful coal reserves, its dwindling oil resources, and the small role oil plays in generating electricity, the issue of whether to rely more heavily on nuclear power has boiled down to a question of nuclear economics versus coal economics. In other countries it may seem wise to build nuclear power plants with a certain disregard for close economic analysis so as to reduce dependence on foreign supplies of oil or coal. But in the United States nuclear power can do little to reduce dependence on foreign oil, and because domestic coal is plentiful, there is little point in replacing coal with nuclear unless nuclear is cheaper. Even spokesmen for the nuclear industry such as Donald Winston, director of media relations for the Atomic Industrial Forum (AIF), have conceded that oil is "almost completely irrelevant" to nuclear economics.

For some years the Atomic Industrial Forum has assembled and published data comparing the costs of coal- and nuclear-generated

---

* There has been serious talk in countries such as Sweden, West Germany, and Russia of using nuclear plants to produce steam for regional heating or industrial processes, but so far, atomic energy has been used exclusively to produce electricity.

electricity, and the AIF surveys consistently have given an edge to atomic power. In a release issued Sept. 16, 1983, for example, the AIF said that "the total cost of producing a nuclear kilowatt hour (kwh) of electricity averaged 3.1 cents in 1982, compared with 3.5 cents for coal." The previous year the AIF reported that nuclear electricity cost an average of 2.7 cents/kwh in 1981 compared to 3.2 cents/kwh for coal. Comparing 1981 and 1982 capacity factors — actual output of electricity as a percentage of the amount plants are theoretically designed to produce — the AIF reported a slight decline in nuclear's performance and a large drop in coal's. Nuclear power plants achieved a capacity factor of 61.0 percent in 1982, the AIF said, while the coal capacity factor dipped to 54.1 percent from 58.6 percent.

Commenting on the 1982 survey, AIF President Carl Walske said that "it would have cost U.S. consumers a whopping $1 billion more . . . if nuclear electricity had been produced by coal." Walske's conclusion followed from the AIF survey results, but not everybody agrees that the surveys adequately capture all relevant data. Commenting on AIF surveys in the February 1983 issue of *Critical Mass,* an anti-nuclear publication, Charles Komanoff compared them to Chicago elections in the days of Mayor Richard Daley — "the suspense in the AIF surveys isn't in the final score but in how the nuclear triumph is rigged."

Komanoff, a consulting economist and longtime critic of the nuclear industry, has complained that AIF surveys are confined to utilities that have at least a 10 percent interest in a nuclear power plant and exclude utilities with no stake in nuclear. As a result, Komanoff charges, the AIF data base "eliminates over two-thirds of the eligible coal-fired capacity, much of which is located in or near coal fields, where coal is cheap and reactors were never built." Writing in the same issue of *Critical Mass,* which Ralph Nader's organization publishes, Komanoff said that the coal plants surveyed by the AIF "paid 0.24 cents/kwh more for fuel than the average national coal plant. . . ."

According to Komanoff the capacity factors that seem to favor nuclear also are misleading. Because nuclear plants cost more to build, he wrote, utilities are reluctant to take them out of service when demand for electricity declines. Instead they take coal plants out of service, which depresses their capacity factors and makes nuclear look better.

In an independent comparison of nuclear and coal costs published by the Union of Concerned Scientists (UCS) in 1983, Komanoff and Eric E. van Loon "found that although both nuclear and coal plant costs grew rapidly in the 1970s, even on an inflation-adjusted basis, nuclear plant costs increased more than twice as fast as coal plant costs." Komanoff and van Loon said that "most of the nuclear increases were due to design and construction changes to improve plant safety, while coal increases resulted from pollution control devices to reduce emissions." The most important factor in the escalation of nuclear costs during the 1970s, they said, was a doubling in the physical quantity of materials — cement and steel, for example — required per nuclear unit. Overall, Komanoff and van Loon found that "nuclear construction now costs over 50 percent more than coal, even when costly air pollution scrubbers are added to coal plants." A study released by the Worldwatch Institute in Washington, D.C., at the end of 1983 came to very similar conclusions. Christopher Flavin, the author of Worldwatch's "Nuclear Power, the Market Test," found that in the last decade "the amount of concrete, piping, and cable used in an average nuclear power plant has more than doubled," while "labor requirements have more than tripled."

Over the years critics of the nuclear industry often have taken the Atomic Industrial Forum to task for ignoring hidden costs and other factors that distort the comparison with coal: the assistance the government has given the nuclear industry in the form of research and development money, personnel training, and fuel-cycle infrastructure; direct subsidies for reactor constructions in the mid-1950s; subsidies from the manufacturers during the early 1960s when they sold plants as "loss leaders"; and the Price-Anderson limit on liability, which applies to nuclear but not coal.

Industry spokesmen respond that the energy sector as a whole is pervaded with government subsidies of one kind or another and that it is absolutely impossible to arrive at clear conclusions about which particular energy industry is helped or hurt most. Industry representatives feel that their critics are motivated by a righteous zeal to abolish nuclear power, not by the spirit of unbiased inquiry.

Comparing 'Typical' Plants

One reason it is difficult to make a dispassionate comparison between coal and nuclear costs is that power plants vary widely, depending on where and when they were built. Both coal and nuclear plants are evolving continuously in response to new regulatory requirements, operating experience, and engineering advances. The only thing everybody agrees on — industry representatives and critics alike — is that nuclear plants are more expensive to build and cheaper to operate than coal plants. But when it comes to determining just how much more expensive or how much cheaper, a great deal hinges on fairly unpredictable variables such as inflation, interest rates, and fuel prices.

Because no single plant can be considered typical of all plants, a midwestern engineering firm — Sargent & Lundy — has devised composite nuclear and coal plants based on the firm's extensive experience as a designer of both nuclear and coal power plants. The Sargent & Lundy study, which the AIF distributed in 1983, draws a comparison between an 1100 MW (megawatt) nuclear plant and two 1100 MW coal-fired generating complexes, one of which would burn low-sulfur coal from the Powder River Basin of Montana and northeastern Wyoming, and the other high-sulfur coal from central Illinois. The nuclear plant would be built in about 12 years, the coal plants in around eight, but all three would be scheduled to come into operation around 1995, and all three would have operating lives of about 30 years.

Sargent & Lundy arrived at conclusions that gave nuclear a very slight advantage over coal. Over its whole lifetime the nuclear power plant would produce electricity at an average cost of 20.8 cents/kwh, the low-sulfur plant at 23.3 cents/kwh and the high sulfur plant at 21.2 cents/kwh.* Construction costs would be about 50 percent higher for the nuclear unit, but fuel costs would be between 2.5-3.5 times higher for the coal plants. Operating and maintenance costs would be about the same for all three.

The Sargent & Lundy estimates are based on a number of assumptions, including projected inflation rates of 7.3 percent to 10.4 percent a year for materials and labor; an average after-tax interest rate of 9.6 percent; and fuel costs ranging from $31/lb. for natural uranium to $36/ton for high-sulfur coal. These assumptions seem reasonable, but it is important to remember that they are merely estimates and that relatively small changes in any of the major variables could drastically affect total generating costs.

Take inflation, for example. If prices for materials and labor rise much faster than expected across-the-board, nuclear costs will be much more adversely affected than coal costs because nuclear plants are more expensive to begin with. Conversely, if inflation is significantly lower

---

* The reason why the low-sulfur plant produces electricity more expensively, apparently, is that higher transportation costs and possibly higher coal costs more than make up for lower pollution control costs.

Electricity Generating Costs
Nuclear vs. Coal—Sargent & Lundy Estimates

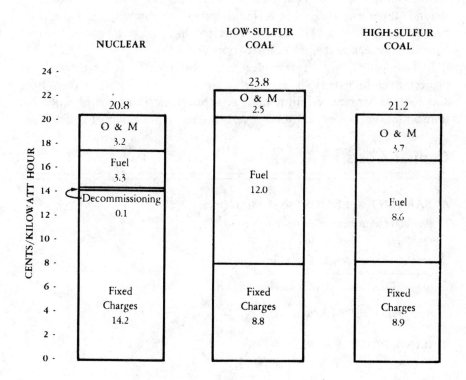

Assumptions: All plants would come into operation in 1994-1995; costs are calculated for a 30-year operating period; the nuclear plant consists of a 1100 MW unit, the coal plants of two 550 MW units each. In the nuclear case, operating and maintenance costs include insurance against nuclear accident and decontamination expenses.

than expected, the final balance will shift decisively in favor of nu-clear. When Sargent & Lundy prepared the study the inflation rate was in fact dropping sharply — from close to 10 percent to around 5 percent.

Interest rates in recent years have been as difficult to predict as inflation. When President Ronald Reagan took office in early 1981 the prime rate was close to 20 percent, and rates have persisted at high levels, contrary to the expectations of many monetary experts. If interest rates remained close to 15 percent over a 15-year period, instead of hovering around 10 percent as Sargent & Lundy postulate, nuclear costs would be much higher. This is because interest rates, like inflation, widen the gap between nuclear and coal construction costs.

Fuel prices, both for nuclear and coal plants, are as difficult to project over long periods as interest rates and inflation. Between 1973 and 1981, as Sargent & Lundy note, coal prices increased sharply and coal transportation prices climbed even higher. Plainly, coal prices are

Capital Costs

A. SARGENT & LUNDY ASSUMPTIONS:

48% Long-term debt at 10.0% interest rate

12% Preferred stock at 10% yield

40% Common stock at 15% yield

100% Capitalization at 12% weighted return

After Tax Weighted Return: 9.6%

B. BANK PRIME RATE: 1979-1983*

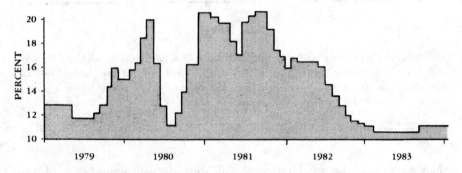

*The prime rate is the interest banks charge the customers they consider most trustworthy. Most customers pay rates well above prime.

Pennsylvania Coal Prices

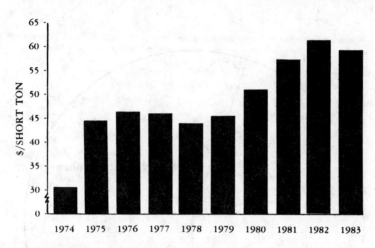

Source: International Monetary Fund

closely linked to oil prices, which increased sevenfold between 1973 and 1981. But oil prices have been volatile, and expert predictions often have been proven wrong in recent years. When oil prices took an upward leap in 1979-1980 top government officials said that the energy crisis was here to stay and warned Americans to tighten their belts. Two years later the oil market was glutted, and energy officials were telling Americans that there would be no more gas lines for the rest of the century.

Sargent & Lundy postulate that high-sulfur coal from Illinois would cost $36/ton over a 30-year period, while low-sulfur coal from Montana would cost $31/ton. One only needs to glance at recent coal quotations, however, to see just how uncertain such assumptions are. With fuel accounting for nearly half the cost of coal-generated electricity, a sharp drop in coal prices would shift the balance decisively against nuclear, while a sustained rise would do the opposite.

Because fuel makes up only about 15 percent of total nuclear generating costs, the final cost is much less sensitive to changes in uranium and uranium fuel-cycle costs. Moreover, because the cost of nuclear fuel contains many independent components, adverse changes in one element may be counterbalanced by favorable changes in others. Spot prices for natural uranium, for example, were running well under

Nuclear Fuel Costs: Sargent & Lundy Estimates

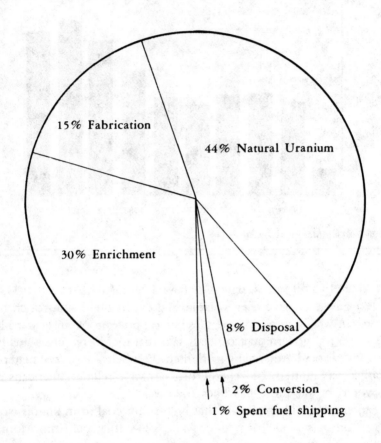

15% Fabrication

44% Natural Uranium

30% Enrichment

8% Disposal

2% Conversion

1% Spent fuel shipping

Based on following assumptions concerning average costs during the 30-year operating life of a nuclear power plant, expressed in 1982 dollars:

Yellowcake	$31.00/UO₂
Conversion of yellowcake to hexaflouride gas	3.10/lb. uranium
Uranium enrichment	130.75/SWU (separation work unit)
Fuel fabrication	
First core	225.00/Kg
Reloads	200.00/Kg
Spent fuel transportation	36.00/Kg
Spent fuel disposal	275.00/Kg

$20/lb. at the end of 1983 — much lower than Sargent & Lundy had postulated. But many critics of nuclear power would argue that the Sargent & Lundy estimates for the cost of spent fuel disposal were much too low.

Plant Delays and Regulatory Reform

Delays in construction, like increases in bank rates or in inflation, usually have a much stronger impact on nuclear costs than on coal. If construction is interrupted or takes much longer than expected, prices will have more time to rise, and the plant's owners will have to pay interest on the money borrowed to finance construction during a longer period. Suppose, for example, that a utility was deciding around 1970 whether to build a nuclear or coal plant for operation in the 1980s. The utility expected the coal plant to cost $500 million to build and the nuclear plant $750 million, or 50 percent more. If, for reasons beyond the control of the utility, the project was delayed for 10 years, prices would have doubled in the meantime. The nuclear plant now would cost $1,500 million to build and the coal plant $1,000 million. The difference in costs between the two plants, which was $250 million in 1970, would be $500 million in 1980.

If the utility opted for nuclear in 1970 and borrowed money to finance the project immediately before the delay, the situation would be even worse. On top of the plant costing twice as much as expected and $500 million more than a coal plant, the utility also would have to pay interest on unused capital over a long period before recovering its investment from rate-payers.

Since the early 1970s lead times for nuclear plants have in fact nearly doubled, and longer construction times have contributed heavily to the escalation of nuclear power plant costs. Critics of the industry usually attribute the delays to shoddy work by nuclear manufacturers and slipshod government supervision, while the industry is inclined to blame interference from citizen activists, fickle regulations, and a cumbersome regulatory process.

Current procedures require issuance of a construction permit by the Nuclear Regulatory Commission (NRC) before work can begin on a nuclear facility. Operating licenses are not issued until around the time construction is completed. License applications are reviewed to determine whether a proposed facility can be built and operated without undue risk to the environment, public safety, and health. The regulatory process often involves lengthy hearings and drawn-out investigations.

The nuclear industry has sought for several years to get the licensing process "streamlined" by consolidating the issuance of construction and operating permits and by limiting the opportunities for citizen intervention. But Congress has not acted on such proposals, and federal regulators deny that their rules are to blame for construction delays. Nunzio Palladino, chairman of the Nuclear Regulatory Commission, pointed out in 1983 that two recently completed plants — St. Lucie 2 in Florida and Palo Verde in Arizona — were "constructed in around half the time most other plants are constructed because [manufacturers] took [NRC regulations] to heart, . . . organized their management and did it."

Officials at the NRC blame many construction delays on slipshod work by engineering firms and reactor manufacturers. One of the most notorious foul-ups that has come to light in recent years involved the Diablo Canyon nuclear plant in southern California, which was built to house Westinghouse pressurized-water reactors. In September 1981, shortly after the plant was licensed to operate, it was discovered that earthquake supports had been reversed so that the strongest parts of the foundation were under the wrong parts of the plant.

Regulators and industry representatives do tend to agree with the desirability of standardizing plant design. The idea is to develop a comprehensive blueprint for plants that could be used universally, expediting engineering and regulatory work. But unforeseen problems in current reactor designs continue to appear, and it may not be possible, as yet, to settle on a single design suitable for use in all situations. As recently as July 1982 the NRC ordered five nuclear plants built by General Electric (GE) to be shut down for inspection of coolant pipes. During the preceding year cracks had been found in 13 other plants built by GE.

Regulatory changes have been especially great since the accident at the Three Mile Island (TMI) nuclear power plant in February 1979 — so great, in fact, that specialists on nuclear economics now distinguish as a matter of course between "pre-TMI" and "post-TMI" generating costs. In a study released by Cambridge Energy Associates in November 1983, I.C. Bupp and Charles Komanoff concluded that electricity from the latest nuclear power plants could be as much as 20 times more expensive than electricity from the first plants. The first plants that entered service by the early 1970s produced electricity for 1 to 3 cents/kwh, Bupp and Komanoff wrote in their study, "Prometheus Bound: Nuclear Power at the Turning Point." The plants that came on stream in the late 1970s generated power for 3 to 6 cents/kwh. But the post-TMI plants that are starting in the 1980s will "produce electricity costing 5-6 cents/kwh, at one end of a wide range, to 18-20 cents/kwh at the other end," they concluded.

Experience and Quality Management

Bupp and Komanoff attributed considerable importance to the experience and skill of utility managers in accounting for the wide variation they found in nuclear generating costs. "Managements of some utilities have coped with the difficult task of building nuclear power plants much more effectively than have others," Bupp and Komanoff observed. "In general, those with the most experience have done best. Clearly, management counts much more heavily in the economics of nuclear power than generally has been believed."

Bupp and Komanoff argued, for example, that reactors have been built more economically in France because a single national utility — Electricite de France (EdF) — is responsible for managing all nuclear projects and a single manufacturer — FRAMATOME — does all the construction work. FRAMATONE can count on receiving four or five orders from EdF every year, and it has been able to organize its procedures so that a fixed number of plants are produced each year in the most efficient possible way. In contrast to countries such as the United States and West Germany, FRAMATOME and EdF carry out their operations "effectively with complete freedom from administrative-regulatory intervention," Bupp and Komanoff noted. According to the Organization for Economic Cooperation and Development, the French have built nuclear plants at an average cost of $680 per kilowatt of installed capacity compared to $1,213 in West Germany and $1,434 in the United States.

In the United States, Commonwealth Edison Company of Illinois and Duke Power Company of North Carolina often are singled out as the utilities that have built nuclear power plants most successfully. Data supplied by Commonwealth Edison indicate that its nuclear plants produced electricity almost twice as cheaply in 1982 as did its coal plants. It paid almost four times as much per kilowatt hour to fuel its coal plants, and even the carrying charges on construction costs were higher for coal than for nuclear.

1982 Coal and Nuclear Generating Costs
Selected Commonwealth Edison Plants (in cents per kilowatt hour)

	COAL AVG*	NUCLEAR AVG*
STATION		
Carrying Charges:		
Plant investment	1.110	.920
Fuel inventory	.222	.261
Expenses:		
Fuel	2.576	.559
Operating & maintenance	.425	.504
Total	4.334	2.244

* Based on Commonwealth Edison data for the Dresden, Quad Cities, and Zion nuclear plants, and the Joliet, Powerton, and Kincaid coal plants. Data are adjusted to a 60 percent capacity factor for all plants.

Critics of the nuclear industry, such as Bupp and Komanoff, concede that Commonwealth Edison achieved a good operating record with nuclear power plants, but they stress that the utility's balance sheets are strongly affected by a number of special factors. First, the nuclear power plants that Commonwealth Edison had operating in 1982 were all built in the late 1960s and early 1970s when reactors were at their cheapest. Second, Commonwealth Edison pays much more for coal than most other utilities.

In January 1984 Commonwealth Edison's management received a severe blow when the Nuclear Regulatory Commission's Atomic Safety and Licensing Board denied it a license to start up the Byron Nuclear Power Station near Rockford, Ill. The Byron plant virtually was completed and had cost $3.35 billion, but the licensing board said that because of poor quality control the safety of the plant could not be guaranteed. It said that Commonwealth Edison had "a long record of non-compliances with NRC requirements."

Slow-Downs and Cancellations

People on all sides of the nuclear controversy usually agree that unexpectedly low demand for electricity has been the factor most responsible for plant delays. Since 1973-1974 demand for electricity has been growing about half as fast as before the oil shock. From the end of World War II up until the OPEC price hikes, demand for electricity grew at an average rate of 7 percent a year. During the last decade the average annual rate has been about 3.5 percent, and in 1982 demand actually dropped by 2.9 percent.

The unexpectedly low rate of demand for electricity has meant that many utilities began to build plants only to discover midway through the projects that the plants would not be needed nearly as soon as expected, or not at all. When nuclear plants are involved the financial penalties have been especially severe. According to Sargent & Lundy, a two-year delay in a reactor project directly after the beginning of engineering work can add $450 million to investment costs. If the delay comes later, after the utility has sunk more money into the project but before it can collect revenues from the sale of electricity, the added costs are even greater.

Generally, state regulatory authorities do not permit utilities to add the costs of a power plant to the rate base until the plant begins to generate electricity. In recent years, though, many financially troubled utilities have sought permission to charge consumers for "construction work in progress." Such requests have become hot political issues in some states and in at least one, New Hampshire, a governor is thought to have been defeated for reelection primarily because of his support for the utility position. The nuclear plant under construction at Seabrook, N.H., is one of the most controversial in the country. The cost of the project, originally estimated at about $1 billion, could go as high as $8-9 billion. In September 1983 the owners of the plant voted to suspend virtually all work on the second of two planned units. Over the next few years, after the first Seabrook unit comes into operation, electricity rates are expected to double in New Hampshire.

The utility group most severely affected by the changing outlook for nuclear electricity has been the Washington Public Power Supply System (WPPSS), a consortium of 88 utilities in Washington state, Oregon, Idaho, and Montana. WPPSS, or "Whoops" as it often is called nowadays, set out to build five nuclear power plants a decade ago. After large cost overruns and long construction delays WPPSS was forced to cancel two plants, and in July 1983 it defaulted on $2.25 billion in bonds.

As the largest issuer of municipal bonds in the country, WPPSS got itself into a situation that threatened to have uniquely far-reaching financial repercussions. But its troubles with nuclear plants were far from unusual. The Public Service Company of Indiana, the Hoosier state's largest utility, declared a "financial emergency" in the summer of 1982 because of delays and cost overruns associated with its Marble Hill nuclear power project, which was started in 1974. Faced with a severe shortage of cash and fearful about how consumers would react to huge rate increases when the plant finally came into operation, the company has proposed a "rate control plan" that would increase electricity prices by 8 percent a year over the next six years. The estimated cost of the Marble Hill plant has climbed to more than $5 billion from $1.4 billion.

A similar escalation in the estimated costs of the Shoreham nuclear power plant built for the Long Island Lighting Company (Lilco) prompted New York Governor Mario M. Cuomo to establish a commission in 1982 to consider the fate of the project. The price of the Shoreham plant, originally estimated to cost $300 million, is expected to come to about $4 billion. In February 1983 the government of Suffolk County, where the plant is located, concluded that it would be impossible to evacuate the area if there were a serious accident. The county refused to adopt an emergency evacuation plan, as required by the NRC, taking the position that the plant never should be opened. By this time Lilco was paying more than $1 million a day in interest on money it had borrowed to build the plant.

When the Shoreham commission issued its report in December 1983 about the only thing the panel's members were able to agree on easily was that the plant should never have been built in the first place. Materials prepared for the commission indicated that Lilco's rates would go up to 28 cents per kilowatt hour from 10.7 cents/kwh by the year 2000 if the Shoreham plant was permitted to open. If the plant is mothballed, rates might climb to about 31 cents/kwh by 2000.

On Long Island and in a great many other regions of the country, a vicious cycle of rate increases and "rate shock" seems to be setting in. The more utilities boost the price of electricity, the harder consumers try to use other sources of energy or to conserve fuels. That in turn lowers demand for electricity and causes the local utility additional financial problems, leading eventually to still higher rates.

In January 1984 the Public Service Company of Indiana abandoned its half-built Marble Hill nuclear plant, which already had cost $2.5 billion. The Cincinnati Gas and Electric Company said it would convert its $1.6 billion Zimmer plant to run on coal. Philadelphia Electric Company announced in January that it would suspend work for 18 months on one of its two nuclear reactors under construction at Limerick, Pa. The delay was expected to raise the cost of the project by $550 million.

Questions in Utility Hearings

The difficulties faced by many nuclear utilities and the prospect of sharp increases in rates have aroused public controversy all over the country, prompting more interest than ever before in the economics of nuclear power. Hearings similar to the ones conducted by Governor Cuomo's commission on the Shoreham plant are becoming commonplace, and as concerned citizens become better informed about the issues at stake, their questions are becoming increasingly sophisticated.

Twenty years ago when utilities first began to order nuclear reactors there was little or no basis for evaluating promises made by manufacturers, engineering firms, utility executives, and public regulators. Nobody had any experience in building, operating, or regulating nuclear power plants. Abrupt changes in energy trends, an everyday feature of life since 1973, were unknown. In hindsight, it is apparent that many of the nuclear projects that were started 10 or 15 years ago should never have gotten off the drawing boards. Now, when no new nuclear plants are being ordered in the United States, the main issue in utility hearings is likely to be whether to finish plants already started or even, in some cases, whether to mothball plants that already are finished.

At the end of 1983, 48 nuclear power plants were under construction in the United States, and nearly half of them were more than three-quarters done. Bupp and Komanoff, in their report for Cambridge Energy Research Associates, considered it "a good bet that most of these partially built units will be finished and enter service before the end of the decade, increasing the nuclear share of U.S. electricity to around 20 percent." It also was a good bet, though, that many of these plants would enter service a great deal later than planned, and in some instances communities would decide to put plants into cold storage even after they were finished.

When it is a question of whether to operate a plant that is already completed, the most important arguments are likely to hinge on how the local utilities, investors, and ratepayers would be affected. Because utilities usually can recover the cost of plants only after they enter operation, the effect of keeping a plant closed often would be to drive a utility into bankruptcy unless special remedial steps were taken. In cases where plants are causing utilities serious financial difficulties, regardless of whether they are opened or not, public officials have to resolve sticky questions about who should bear the brunt of added costs. Consumers naturally resent having to pay higher rates because of mistaken decisions that they had little or no part in taking or that they even may have opposed. Investors usually feel that they should not have to pay a penalty, having accepted relatively low rates of return in exchange for putting their money in what was supposed to be safe investments. Utility executives often argue that they also are blameless because their decisions were approved by public regulators at every stage of the way.

In a country such as France, where all plants are built by a single national company, the costs of building an unneeded plant naturally can be divided up among ratepayers and taxpayers all over the country. This kind of solution is of course not available in the United States, but in instances where local utilities have gotten themselves and their communities into disastrous difficulties it may be necessary to find some means of getting help from larger jurisdictions. Most Americans are sure to feel that the local communities, their elected officials, and their utilities should bear the brunt of their own errors. But at a time when the same kinds of errors — mistaken energy projections and overly optimistic expectations about building costs — have been made by utility executives, manufacturers, and regulators, there may be a case for getting everybody to contribute something to help alleviate the problems of the most unfortunate. There may even be a case for some federal aid, considering that the U.S. government did much for several decades to encourage utilities to "go nuclear."

In cases where utilities are torn about whether to complete nuclear plants that already are almost finished, the issues are essentially similar to the questions that come up when it is being decided whether to build a facility in the first place. Anticipated changes in electricity demand have to be evaluated in terms of local economic trends and possibilities for conservation and global fuel prices. When alternative generating capacity is available, it is important to take a critical look at the presumed costs of using it. Utility executives and regulators may believe that it would be more expensive to rely on existing coal capacity, for example, than to finish and open a nuclear plant. But their position will depend on assumptions about projected coal prices, inflation, and interest rates. In cases similar to Shoreham, where it is a close call whether to complete and operate a nuclear plant, communities may decide — as Suffolk County did — that safety considerations tip the balance against nuclear.

Many economic variables will remain elusive and intangible no matter how hard one tries to grasp nuclear economics. Most plants under construction in the late 1970s and early 1980s have begun to operate and it remains to be seen whether they will stand up as well as they are supposed to during their whole projected lifetimes. The industry still has no firm plan for final disposal of spent fuel from reactors, and the costs of disposal could turn out to be much higher than the industry and regulators currently expect. Utilities have had very little experience decommissioning and dismantling plants, and this, too, could bring on unforeseen difficulties and costs. Finally, there is always the danger of a major nuclear accident, which could involve enormous added costs: lost investment money, cleanup expenses and — not least — compensation for damage to property and people.

Because the hazards associated with nuclear accidents and waste disposal are so serious, people are inclined to worry mainly about the hazards themselves and only secondarily about their economic dimension. But their economic dimension is important too, and it should not be forgotten when the costs of nuclear and coal are being tallied up.

Three of the four cooling towers at the Three Mile Island nuclear power plant near Harrisburg, Pa. On March 28, 1979, a malfunction in the plant's reactor cooling system caused one of the worst nuclear accidents in the history of commercial nuclear energy.

Chapter 5

UNCERTAIN RISKS AND UNSOLVED PROBLEMS

The nuclear controversy often is described as a conflict between values and social visions as well as an argument over economics and safety. In truth, proponents and opponents of nuclear power sometimes talk as though the whole shape of society will depend on whether atomic power plants are built or not. On the one side, people claim that the nation will sink into a position of economic inferiority marked by electricity "brownouts" and industrial stagnation if construction of power plants is not accelerated. On the other side, people warn that the whole "global village" will run out of diminishing fuel supplies unless we place more emphasis on development of renewable resources such as wind, water, and the sun.

Admirers of E. F. Shumacher, the British economist who wrote *Small Is Beautiful*, claim that large energy systems are bad for self-reliance, political liberty, community spirit, and innovation. They would prefer to live in a society in which most energy is produced locally from a variety of renewable sources. Critics doubt, though, whether it would ever be possible for highly urbanized and industrialized societies to rely heavily on decentralized energy systems. In the nuclear industry, calls for decentralized systems are almost always dismissed as utopian, and industry representatives often attribute the hostility they meet among environmentalists and community activists to irrational fears and factual misunderstandings.

Engineers connected with the industry like to remind their critics that almost every major technological innovation — household electricity, for example, or commercial aviation — aroused intense public anxieties when it was first introduced. Nuclear power plants have encountered especially intense hostility, it is felt within the industry, because of their association with the mushroom cloud. Engineers and executives often have expressed confidence that the public will relax about nuclear power once it is universally appreciated that atomic plants cannot explode like atomic bombs.*

It surely is true that the industry has suffered from unpleasant mental associations, and it undoubtedly also is true that new technologies encounter resistance until members of the public have grasped the risks involved and learned to deal with them. In the case of nuclear energy, however, the industry's situation has been enormously aggravated by the fact that some risks still are not fully understood and some crucial problems still remain unsolved. The industry started out at a big disadvantage because dangers connected with radioactive materials were not understood or appreciated in the early years of nuclear research. Many people suffered as a result, and when atomic power plants were first built much of the public was predisposed to wonder whether the technology was as safe and problem-free as the industry claimed. In recent years mishaps and misfortunes in the industry and the federal regulatory apparatus have seemed to bear out the public's suspicions.

Assessing Radiation Risks

From the dawn of the nuclear age public fears have centered on radiation, an insidious and invisible killer that can penetrate biological tissue and cause cancers or genetic defects. X-rays used for medical diagnoses are one form of radiation, and in the first decades of the 20th century many patients contracted cancers after being exposed recklessly to massive doses of X-rays.

Physicists and chemists working with radioactive materials in the early years also appear to have suffered from radiation exposure, although it is impossible to prove beyond doubt that radiation was the main cause or only cause of diseases. Marie Curie, the first person to refine radioactive substances such as radium, died of leukemia. So did the daughter of Marie and Pierre Curie, Irene, who pioneered the production of artificial radioactive substances with her husband, Frederic Joliot-Curie.

The kind of radiation emitted by radioactive substances is called ionizing radiation, in contrast to the less powerful radiation that comes from high-voltage power lines, radar installations, and microwave towers.

---

* Hydrogen gas or steam could cause an explosion in a commercial reactor, but the explosion likely would only be strong enough to rupture the container vessel. The danger would lie not in the explosion itself, but in the subsequent release of radiation.

Ionizing radiation has enough energy to knock electrons loose in biological tissue, damaging cell structure and causing chemical reactions. Ionizing radiation can prevent a cell from dividing, or it can damage the genetic materials contained in the cell's nucleus, making it divide abnormally.

It was plain long before atomic power plants were built that radiation was and is profoundly dangerous to human health, but discovering exactly how dangerous has proved to be a difficult enterprise that is far from finished today. There are several different kinds of radiation — alpha particles, beta particles, gamma rays, and neutrons — and they affect biological tissue differently. Neutrons and alpha particles are weak penetrators, but if they enter an organism by inhalation, ingestion, or through an abrasion, they emit a dense trail of ionization and are intensely harmful. Beta particles and gamma rays are less harmful but much more penetrating.

The way a body reacts to radiation depends on what tissue is affected as well as the type of radiation involved. Rapidly dividing cells, such as the cells found in the human fetus, are exceptionally vulnerable. So are certain organs that tend to attract and concentrate radioactive substances. Strontium-90 migrates to the bones, for example, where it can emit radiation over a long period of time. Plutonium concentrates in the liver and bones and secondarily in the gonads. To further complicate matters, some age groups are especially susceptible to certain cancers that radiation can cause. Young children, for example, are more likely to contract leukemia.

Most cancers induced by exposure to radiation appear only after a long delay, and genetic defects become evident only after exposed people have produced offspring. Injuries generally will show up only in a tiny portion of the population that is exposed or "at risk." Consequently, assessment of radiation risks requires statisticians and health physicists to monitor large populations over long periods of time, which usually involves daunting and sometimes insuperable logistics problems. As defects appear in a population or in the population's successor generations, they must be correlated with data collected on how much radiation and what kind of radiation individuals in the population were exposed to.

Serious uncertainties arise if the data on radiation exposure — the "dosimetry" — turn out to be incomplete or unreliable. Especially problematic questions of interpretation come up if researchers discover that individuals or whole populations have been exposed to cancer-

73

causing dangers other than radiation such as toxic chemicals. Populations also may have been exposed to unusually high levels of natural radiation. In some parts of the western United States, high levels of cosmic rays greatly increase the amount of natural radiation a person absorbs as a matter of course; in Kerala, a state in southern India, natural deposits of radioactive minerals are so large that they may explain exceptionally high rates of genetic defects such as mental retardation.

Before World War II the numbers of people exposed to significant amounts of artificial radiation were rarely large enough to permit really meaningful epidemiological studies (studies on the incidence of health defects in populations). Since the war, however, more elaborate statistical studies have been conducted on a number of special groups that came into contact with ionizing radiation for one reason or another. The groups include the survivors of the Hiroshima and Nagasaki atomic bombings; people who entered the two cities right after the war when radiation levels were still high; American soldiers who conducted military maneuvers during atomic bomb tests in the 1950s; civilian populations living near test sites; workers in the U.S. facilities that produce radioactive material for atomic bombs such as the Hanford Nuclear Reservation in Washington state; and workers who repair nuclear submarines for the U.S. Navy. Studies of groups exposed to X-rays for diagnostic or therapeutic purposes also have been relevant to the general assessment of radiation risks.

Interpretation of such studies has proved to be an extremely arduous and controversial business, which has divided statisticians and health physicists into feuding clans, each determined to defend and advance distinct points of view. Depending on who prevails in the end and who actually is right, lives could be needlessly lost or nuclear activities could be unnecessarily curtailed by such interpretations. Both in the nuclear industry and among the scientists who consider themselves the guardians of nuclear workers and public health, the controversy over radiation hazards is seen as a matter of life or death.

Major Study Results

The Atomic Industrial Forum (AIF), which ordinarily is careful to argue from positions that generally are supported in the scientific community, made an uncharacteristic slip when it claimed in an April 1983 public information bulletin that radiation is "the most scientifically

Statistical Theories About Incidence of Cancer

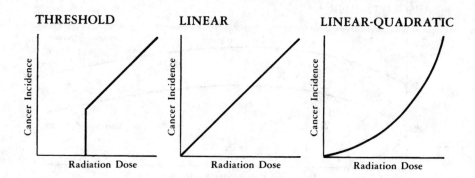

In the threshold theory, there is no incidence of cancer at low doses, but at a certain level cancer incidence increases with dose. In the linear theory, with every increase in the amount of radiation exposure there will be a proportionate increase. In the linear-quadratic theory, cancer incidence is relatively rare at low levels of exposure but becomes increasingly common at higher doses.

understood, easily detected and readily controlled of all environmental agents." [1] The AIF also departed from mainstream scientific opinion when it said in the same release that radiation can cause cancers at high doses but probably not at lower doses "below about 10,000 millirems." *

In point of fact, the theory that radiation is only dangerous above some threshold has been losing ground among scientists in recent years. The more generally accepted "linear theory" holds that radiation is dangerous at all levels of exposure and becomes steadily more dangerous as the level rises. The threshold theory continues to have passionate adherents, however, and they are likely to defend their position for years to come.

A similar dispute has been raging between supporters of the "absolute risk" and "relative risk" models. In the absolute risk model, ra-

---

* A millirem is one thousandth of a rem, the standard measure of radiation's biological impact. The rem is derived by multiplying a "quality factor" based on the type of radiation involved times the quantity of radiation as measured in rads. The rad is a precise unit defined as the quantity of radiation that will cause one kilogram of material to absorb 0.01 joules of energy. On average, an American receives about 110 millirem of radiation from cosmic rays each year.

Absolute and Relative Risk Models

ABSOLUTE RISK

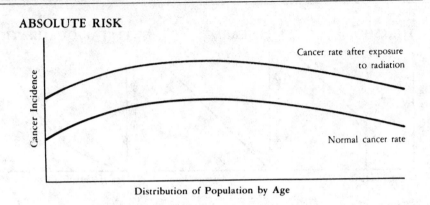

Cancer rate after exposure to radiation

Cancer Incidence

Normal cancer rate

Distribution of Population by Age

RELATIVE RISK

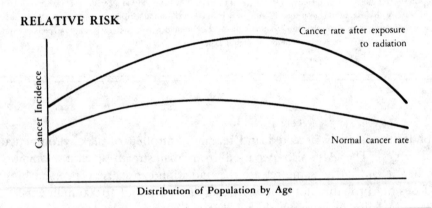

Cancer rate after exposure to radiation

Cancer Incidence

Normal cancer rate

Distribution of Population by Age

diation increases the percentage of people who will contract cancer in an exposed population, while the relative risk model postulates that radiation causes a percentage increase in the proportion of people who would have contracted cancer anyway.

Scientists trying to sort out such questions have found the results from the major epidemiological studies to be puzzling, inconsistent, and contradictory. In some cases, studies have given rise to charges and countercharges about alleged political interference in the awarding of contracts. Individual scientists, on occasion, have changed their minds quite drastically about their own data. In the latter years of the Carter administration there was considerable pressure to remove radiation research from the Department of Energy's (DOE's) jurisdiction and to centralize all research in the Health and Human Services Department.

As of February 1984 the most important and meticulous research by far had been devoted to the survivors of the Hiroshima and Nagasaki nuclear bombings. Collection of data and materials began in 1945 and was consolidated in 1947 under the aegis of the U.S.-created Atomic Bomb Casualty Commission. In 1975 the commission was replaced by the Radiation Effects Research Foundation, which is managed jointly by Japan and the United States. Researchers found a very elevated incidence of leukemia during the first decades after the bombing and a significantly higher incidence of thyroid, breast, and lung cancers. Malignant tumors appeared at abnormally high rates after an average latency period of about 20 years, and in 1982 researchers reported that the incidence is still rising as the survivor population ages.

Surprisingly, death rates from non-malignant causes have been lower than average among survivors, and children exposed to radiation in the womb have not suffered an abnormally high incidence of leukemia. Most startling of all, no significant genetic defects have been found in the survivor population, even though it is well established that radiation can cause genetic damage. Some researchers believe that the results show radiation to be less dangerous than previously thought, but others argue that radiation actually may be more dangerous.

Dr. Josef Rotblat, a former president of the British Institute of Radiology, resigned from the Manhattan Project at the end of World War II when it was becoming apparent that the atomic bomb would be used against Japanese cities. He has argued for some years that the Hiroshima and Nagasaki survivors are a "selected population" in which "only the genetically toughest individuals survived the injuries and trauma of nuclear attack." [2] Writing in 1977, Rotblat suggested that the survivors of the bombings "may not be suitable as the basis for calculating radiation risk factors for other populations, since the survivors may have been exceptionally resistant to both cancers and genetic damage." [3]

Further doubts were cast on the meaning of the Japanese results in 1982 when the original estimates of the neutron doses victims of the Hiroshima bombing received were found to be too high, implying that cancers were caused by lower levels of radiation than previously thought. New estimates are being prepared in the United States and Japan.

Meanwhile, radiation specialists are divided about what the ultimate implications will be.

Dr. Alice Stewart, a leading British epidemiologist, has said that ionizing radiation should be considered about 10-20 times more dangerous than is assumed by current radiation standards. Seymour Jablon, a radiation expert with the U.S. National Academy of Sciences, thinks that radiation will be judged at least twice as risky. Others insist that the prevailing standards are about right. Since 1972 the National Academy of Sciences' Advisory Committee on the Biological Effects of Ionizing Radiation has made three reports — BEIR I, II, and III — without being able to arrive at any firm conclusions that all participants were willing to endorse.

Radiation Risks to Workers

With scientists unable to agree on the effects of low-level exposures to radiation, it has been difficult for the public to feel sure that radiation releases from nuclear power plants are as insignificant as regulators and industry representatives usually say they are. Scientists who make alarmist statements about radiation releases from plants can count, generally, on getting sympathetic attention in the press. Lacking a yardstick against which allegations can be measured, editors and reporters have seen no reason to ignore persistent critics such as Dr. Ernest J. Sternglass of the University of Pittsburgh who has been attacking the industry's radiation standards for years.

Nevertheless, Dr. Sternglass has been almost unique in maintaining that radiation releases from atomic power plants are great enough to be a significant hazard to public health. Scientists generally agree that reactors present little danger to the public, as long as they are working properly. Whether reactors and the facilities needed to support their operations present significant dangers to workers is a somewhat more controversial matter.

Dr. Bernard L. Cohen, a prominent defender of nuclear power at the University of Pittsburgh, wrote in 1979 that cancer risks attributable to radiation reduce the average nuclear worker's life expectancy by 20 days, while occupational accidents cost the average worker in all other industries nearly four times as many days.[4] Dr. Cohen observed that the average nuclear worker is exposed to just 0.7 rem a year, while the legal limit on such exposure is seven times higher — 5 rem.

In a rejoinder to Cohen, Dr. Karl Z. Morgan noted that thousands of individuals in the industry receive more than 0.7 rem a year. Dr. Morgan, a former chairman of the International Commission on Radiological Protection and the National Council on Radiation Protection, said it "is little consolation to the worker dying of a radiation induced cancer to know that the average employee received 1/7 his exposure." [5]

Dr. Morgan is among those who believe that exposure standards for workers should be considerably tightened because of the accumulating evidence that risks associated with low radiation doses have been underestimated. But Dr. Cohen in his 1979 article referred to this contention as "a standard weapon in the arsenal of those dedicated to the destruction of nuclear energy." Cohen claimed that reducing the exposure limit by a factor of 10 would cost the nuclear industry $500 million and save, at most, 10 lives. He said "it would be difficult to justify spending $500 million to save 10 lives when our society can save a life for every $25,000 spent on medical screening programs and for every $100,000 spent on highway or automobile safety devices."

Dr. Cohen and like-minded advocates of nuclear energy concede that risks vary substantially within the industry depending on what workers do. People who mined uranium in the early years of atomic power ended up contracting lung cancer in staggering numbers. This tragedy could have been averted if mines had been properly ventilated, and it was known at the time that ventilation was necessary. Even today workers in uranium mining, reactor construction, and certain fuel-cycle facilities such as the British and French reprocesssing plants face higher risks than workers at atomic power plants.

Those who seek stricter radiation standards have been particularly concerned about the industry's practice of employing temporary workers for "hot jobs" in which individuals might be exposed to their yearly radiation limit in just a few minutes. The practice is considered acceptable under current standards, but critics find it difficult to believe that it makes no difference whether one is exposed to the limit in a year or a minute — it sounds too much like saying that one could just as well drink one martini a night for a year or 365 martinis in one evening.

The extent to which the industry will have to keep using "jumpers" on hot jobs may depend greatly on how well equipment turns out to work as it ages. If reactor pipes, remote control equipment, and fuel disposal tanks hold up well, it should not be necessary for workers to be exposed to intense doses of radiation except under rare circumstances. If, on the other hand, repairs increasingly are required, the dangers facing workers could be higher than expected.

Catastrophic Reactor Accident

Ever since nuclear power plants were first built in the late 1950s, reactor specialists have been concerned about what the consequences would be if a reactor's cooling system failed and the fuel melted, burning through the foundation of the plant into the earth and groundwater below. The number of diseases and fatalities that might occur during the decades after such an accident varies considerably with alternative assumptions about radiation hazards. Under any assumption, however, a reactor meltdown — the so-called "China syndrome" — would be a catastrophe of terrible proportions.

In the first major government report on reactor accidents, which was prepared for the Atomic Energy Commission (AEC) in 1957, the research team considered the possibility of a meltdown at a power plant located 30 miles from a large city. In the worst case, the team concluded, 3,400 people would die, 43,000 would be injured and there would be $7 billion in property damage.

In 1964 scientists at Brookhaven National Laboratory on Long Island did an update of the 1957 report for the AEC. The scientists found most of the assumptions and the methods used by the 1957 team to be sound. But largely because of the fact that the average size of nuclear power plants had increased dramatically in the meantime, it was necessary to raise estimates of how many people would be affected by an accident. After considering the possibility of a meltdown at a 1000 MW plant, rather than the 150 MW reactor the 1957 team had studied, the Brookhaven scientists concluded that an accident could result in 45,000 deaths. They found that the area affected by such an accident "might be equal to that of the state of Pennsylvania."

Fearful about how news of such a report might affect public opinion, the AEC decided to bar publication of the 1964 report and to keep its results secret. In 1972, after people from the Union of Concerned Scientists drew attention to the meltdown problem,* the AEC's chairman, James Schlesinger, commissioned a new reactor safety study. Schlesinger hoped, apparently, that such a study would vindicate reactor

---

* See Chapter 2.

safety and lay public concerns to rest. The contract to direct the preparation of the study went to a nuclear engineer at the Massachusetts Institute of Technology named Norman C. Rasmussen.

The Rasmussen report, known as WASH-1400, was released in 14 volumes in August 1974. On the basis of elaborate theoretical models designed to take into account the possible ways an accident could occur, the report said that "a person has about as much chance of dying from an atomic reactor accident as of being struck by lightning." Critics considered this conclusion, which was highlighted in most news reports on the Rasmussen study, to be highly dubious considering the vast number of uncertainties involved. During the following years, the study came under attack from many scientists and anti-nuclear activists. Finally, in January 1979 the Nuclear Regulatory Commission (NRC) — the successor to the AEC — withdrew official endorsement of the study, saying that it no longer regarded the study's numerical estimate of accident risks as reliable.

The NRC acted in the nick of time in terms of public relations because just two months after it repudiated the Rasmussen report's key conclusion a major accident occurred at the Three Mile Island Nuclear Station near Harrisburg, Pa. The accident resulted from a combination of equipment failure and human error: A valve stuck open so that water drained from the reactor core; the operators, not realizing the valve was open, turned off the emergency cooling pumps that had come into operation automatically; the reactor rapidly overheated and, by the time operators realized something was wrong, materials in the core had started to melt. Later studies indicated that the reactor came within a half hour to an hour of a complete meltdown.

Over a period of days the country's attention was riveted on the Three Mile Island plant as an emergency team from the NRC worked frantically with utility engineers to bring the reactor under control. In news reports from the scene, it was clear that the experts were encountering problems they were not completely confident they could solve. It was apparent, in fact, that they were not even sure they understood what was happening in the reactor.

The most dramatic unexpected problem was a buildup of hydrogen, a highly explosive gas, in the reactor dome. While this problem had been mentioned as a possibility in the literature on reactor safety, it had never been greatly emphasized, and it brought home the fact that even if nuclear power plants could not explode like atomic bombs, they could conceivably explode. In this particular case, the NRC's specialists believed at the time that an explosion was an imminent possibility unless preventive action was taken. Later studies indicated that there was no serious danger of an explosion.

Three Mile Island, or "TMI" as it became known in the daily press, was not the first serious accident or near miss in the United States. In 1962 an experimental breeder reactor partially melted down near Detroit. In 1975 a fire broke out under the control room at a power plant at Browns Ferry, Ala., where technicians were doing work on electrical cables. As a result, operators lost control of the plant's cooling systems and were barely able to get the reactor turned off.

Among specialists on nuclear energy the accident at Browns Ferry made a big impression because it showed that something can always go wrong no matter how many backup systems and fail-safe mechanisms are built into a reactor. But for most of the public this lesson became apparent only with the accident at Three Mile Island. TMI seemed to illustrate the old truth that given enough time, anything that can go wrong will go wrong. It showed that a nuclear disaster is not merely an abstract and remote possibility but something that can happen any time at any of the plants operating around the country.

Implications of TMI Accident

Everybody agrees that a serious nuclear accident must now be considered a real possibility, but beyond that opinions vary widely — even wildly — on the meaning of TMI. To some specialists on reactor safety, TMI confirmed arguments that nuclear technology is fundamentally unsound and never will be foolproof enough to satisfy reasonable safety standards. To others, TMI showed that even when a great deal goes wrong, reactors are not likely to slip entirely out of control.

Daniel Ford, who was the first to call public attention to the reactor meltdown issue as a staff member with the Union of Concerned Scientists, concluded from TMI that the government should retreat from its position that nuclear energy is essential to the country's economic security and progress. Writing in *The New Yorker* on Nov. 1, 1982, Ford said one could "only hope that before the next nuclear accident the authorities will finally recognize that when one is at the edge of a precipice the only progressive move is a step backward."

For many involved in the anti-nuclear movement, Ford's argument for a retreat is considered too moderate a response to TMI; many critics of nuclear energy would like to shut down all nuclear reactors immediately. The Atomic Industrial Forum, on the other hand, has argued that TMI did not fundamentally change what is on the whole a nearly unblemished safety record. "No member of the public has been injured or killed from a reactor accident at a commercial nuclear power plant," the AIF said in an April 1983 bulletin. "No plant employee ever has exhibited clinical evidence of serious injury from radiation," the AIF continued, and "the nation's most serious commercial nuclear plant accident . . . did not alter this unparalleled record of safety."

A 12-member commission established by President Carter to evaluate the implications of TMI concluded that the accident resulted from compounded human failures on the part of the utility, its employees, and federal regulators, as well as design deficiencies and a failure to learn from earlier incidents. The commission, which was headed by John Kemeny, then president of Dartmouth College, said that "training of the operators was totally inadequate for accident conditions." The commission thought "that perhaps the underlying cause [of this failure] is that a great many previous incidents and warnings that had been issued were ignored both by the utility and by the Nuclear Regulatory Commission."

The Kemeny Commission's report said that "if the country wishes . . . to confront the risks that are inherently associated with nuclear power, fundamental changes are necessary if those risks are to be kept within tolerable limits." The commission recommended, among other things, the abolition of the NRC and its replacement by a new executive agency headed by a single administrator rather than a board of commissioners.

Since TMI the NRC has gone through a change of leadership, but critics say that the commission still worries too much about defending the long-term interests of the nuclear industry and too little about the safety problems it is supposed to be solving. In October 1983 an internal government report was released that said the public was justifiably skeptical about the NRC's work. The report, which was prepared by Helen H. Hoyt, an administrative law judge connected with the Atomic Safety and Licensing Board, recommended the creation of an independent inspector general within the agency. Judge Hoyt also recommended establishment of a hot line for "whistleblowers," more training for plant inspectors, and more thorough NRC investigations so that criminal charges could be brought when necessary against violators of regulations.

Whatever else one may say, the NRC has tried harder since TMI to project a hard-nosed image of itself, and its decisions have been costly for the industry it supposedly is coddling. The commission has imposed large fines on plant operators on a number of occasions; it has ordered numerous plants shut down because of safety defects; and it has come into sharp conflict with local authorities over the adequacy of evacuation plans, threatening for example to close down the Indian Point nuclear plant north of New York City unless the local county improved emergency planning promptly. The design changes the NRC has imposed on builders of nuclear plants are so extensive that specialists on nuclear economics now distinguish, as a matter of course, between pre-TMI and post-TMI power plant costs.

No doubt arguments will continue to rage about the technical implications of TMI and the regulatory response. One thing that already is perfectly plain, though, is that the accident has been extremely costly for the utility, ratepayers, and taxpayers. General Public Utilities (GPU), the owner of the plant, filed claims for $4 billion in damages against the NRC and Babcock & Wilcox, the manufacturer of the TMI 2 reactor. While the suits were unsuccessful, $4 billion apparently represents an estimate — perhaps inflated — of what the accident will cost GPU taking all losses into account: sunk construction costs, lost revenues from electricity sales, and cleanup expenses.

The cleanup costs alone are expected to come to about $1 billion — about twice the cost of constructing the TMI 2 reactor. To help GPU avoid bankruptcy, the federal government, the state of Pennsylvania, and private utilities around the country have agreed to kick in money for the cleanup. GPU's customers are paying about $36 million a year in higher rates to help cover the cleanup costs.

The cleanup is an unprecedented operation, which involves the decontamination, removal, and disposal of highly radioactive materials. The operation did not even begin until September 1981, two and a half years after the accident when the plant had cooled down and levels of radioactivity had dropped somewhat. Workers were not able to get a look at the damaged core until summer 1982. In early 1983 a number of engineers connected with the cleanup operation resigned, charging that the effort was riddled with mismanagement, inefficiency, and waste. The following September, NRC investigators concluded that the companies responsible for the cleanup had circumvented proper procedures and adopted methods of questionable safety.

The TMI cleanup operation has highlighted the fact that utilities have had virtually no experience, to date, with dismantling commercial power plants and disposing of their components. Many small experimental reactors have been taken out of service, but only two have been broken up and carted away. Not one full-scale commercial plant has been dismantled, and nobody knows just how difficult the process will turn out to be. Work on dismantling the Shippingport reactor, the nation's first nuclear power plant, has been scheduled to begin in 1984.

Unsolved Waste Disposal Problems

Uncertainties surrounding the dismantling and disposal of nuclear power plants are symptomatic of a situation in which an entire industry was built before anybody knew exactly how its waste products would be disposed of.

Basically, the nuclear industry produces three types of waste: uranium mill tailings, which are a byproduct of uranium mining; low-level wastes, which are materials and equipment that have been slightly contaminated in reactor and fuel-cycle operations; and spent fuel. If spent fuel is reprocessed — that is, if plutonium and uranium are separated out for reuse in reactors or for use in atomic bombs — the extremely radioactive residuals are known as "high-level wastes."

Uranium mill tailings and low-level wastes present relatively straightforward disposal problems, but in the early years of the nuclear era these problems were not always solved responsibly. Low-level wastes simply were dropped in the ocean, for example, with little concern about how they might affect marine life. Mill tailings, which contain about 85 percent of the radioactivity contained in the original uranium ore, were used widely in the western states as construction materials. They may have been responsible for many unnecessary diseases and deaths.

Despite such scandals, the disposal problems connected with mill tailings and low-level wastes generally are considered to be readily solvable within the framework of standard technical procedures and institutional arrangements. The federal government began to take effective action on uranium mill tailings in 1978 after some 140 tons had piled up at active and inactive mills. Much of the material was at poorly secured sites where it was susceptible to wind and water erosion. The federal Uranium Mill Tailings Radiation Control Act of 1978 established a remedial program to stabilize the sites, with the U.S. government covering 90 percent of the costs and state governments the remaining 10 percent. The act strengthened cooperative arrangements between the NRC and state regulators, and it charged the U.S. Environmental Protection Agency and the NRC with setting and enforcing standards.

As of February 1984 more than 175 million tons of mill tailings had accumulated at 27 sites in the United States and they were continuing to pile up at a rate of 10-15 million tons a year. In late 1983 the Department of Energy launched a three-year investigation to survey some 8,000 other spots where tailings may have been used as construction materials. These locations are in Arizona, Colorado, Idaho, North Dakota, Oregon, South Dakota, Texas, Utah, and Wyoming.

While the amount of radiation that tailings emit at any one time is quite small, the radioactive substances in the tailings are long-lived — meaning that they will continue to emit radiation over a long period of time. Over a period of 100,000 years, the cumulative amount of radiation emitted by the tailings "becomes the dominant contribution to radiation exposure from the nuclear fuel cycle," NRC Commissioner Victor Gilinsky pointed out in 1978.

Low-level wastes are problematic mainly because of their huge volume. EPA estimates that there will be about one billion cubic feet of low-level wastes in the United States by the year 2000. Already, land disposal sites are filling up, reports of illegal dumping are heard, and the few states that operate disposal sites are becoming increasingly reluctant to accept wastes from other states. The U.S. Low Level Waste Policy Act of 1980 made the disposal of low-level nuclear waste a state responsibility subject to existing federal regulation.

Apart from the defense sector, roughly 40-45 percent of the country's low-level wastes come from the nuclear energy industry. About a quarter comes from industries such as radiopharmaceutical manufacturers and another quarter from medical and research centers. About 5-10 percent comes from non-defense-related government activities. In addition the government's nuclear weapons production facilities produce a large volume of low-level wastes and a very large volume of high-level wastes. The reprocessing plants that produce plutonium for nuclear weapons are in fact the most important single source of high-level wastes.

High-level wastes and spent fuel, in contrast to mill tailings and low-level wastes, present formidable technical and institutional problems.

They are intensely radioactive, and some of their constituent substances — the long-lived transuranic elements such as Americium 241 — will remain dangerously radioactive for thousands of years. While the physical quantity of high-level wastes and spent fuel assemblies is relatively small, the technical problems connected with their permanent disposal have yet to be satisfactorily solved. Critics of the industry doubt whether there is any satisfactory method of storing the wastes so that they can be counted on to remain intact longer than any human institution has survived up until now.

High-level wastes have been generated almost exclusively by the giant reactors at Hanford, Wash., and Savannah River, S.C., where all the plutonium for U.S. nuclear weapons is produced. As of February 1984 more than 75 million gallons of such waste had accumulated at the two installations, where it was stored temporarily in steel tanks. Between 1956 and 1976 some 500,000 gallons leaked into the ground at Hanford before it was discovered that the tanks had corroded faster than expected.

The Department of Energy, as the direct descendant of the part of the AEC responsible for research and development, manages the facilities where materials for nuclear weapons are produced. DOE also manages the government's nuclear research facilities such as Idaho Falls, Idaho, where more than three million gallons of high-level wastes are stored. DOE is responsible for virtually all work connected with the permanent disposal of wastes from both military and commercial facilities. The NRC has oversight authority, however, over any DOE facility built to accommodate commercial wastes.

At one time it generally was assumed that spent fuel from commercial power plants would be reprocessed, yielding recyclable plutonium and residual high-level wastes that would be disposed of together with high-level wastes from military facilities. But in recent years it has become apparent that the cost of separating plutonium from spent fuel would exceed the price of buying new uranium.* At the same time a growing body of scientific opinion has come around to the view that reprocessing may complicate the waste disposal problem about as much as it helps it. Reprocessing reduces the volume of wastes that have to be disposed of in the most careful possible way, but it does so at the cost of generating additional low-level wastes, which also must be stored.[6] As of

---

* See Chapter 2.

early 1984 government planning for the disposal of nuclear wastes assumed that it will be stored as spent fuel, without reprocessing.

Between 1954 and 1980 some 25,000 spent fuel assemblies from commercial power plants accumulated, and facilities for temporary storage of the assemblies adjacent to power plants began to fill up. With spent fuel assemblies piling up at an increasingly fast rate, utilities have pressed the government to establish "away from reactor" facilities for temporary storage of assemblies. But environmentalists and anti-nuclear activists successfully have blocked the proposal. Critics of the nuclear industry thought that creation of new temporary facilities would unnecessarily add to the amount of radioactive material being transported around the country and divert attention from the urgent need to come up with a means of providing permanent storage.

Congress, increasingly alarmed by the inability of the federal government, state governments, the nuclear industry, and environmentalists to arrive at a consensus about waste disposal, enacted the Nuclear Waste Policy Act in 1982. The law, which President Reagan signed on Jan. 7, 1983, sets up a timetable for establishment of a permanent, underground repository. DOE is to choose five possible sites for such a repository and recommend three to the president by Jan. 1, 1985; by 1987 the president is to recommend the first repository site to Congress; and the site, after being prepared, is to begin receiving nuclear waste no later than Jan. 31, 1998.

Unfortunately, even the U.S. Congress cannot ensure that technical problems will be resolved by 1998. The first site the government selected for study — a salt dome in Lyons, Kans. — was abandoned in 1972 when scientists came to the conclusion that it might contain too many mining boreholes. They were afraid that moisture might leak into the repository and carry radioactive materials into the groundwater. A second site, located near Carlsbad, N.M., began to look like a dubious choice in fall 1982 when a pocket of brine was discovered under the salt bed. Anticipating such problems in 1978, an inter-agency task force on nuclear waste disposal recommended looking into non-salt formations such as basalt and granite. At the beginning of 1984 DOE was evaluating geologic zones at Hanford, Wash.; Beatty, Nev.; David Canyon and Lavender Canyon in Utah; Swisher County and Deaf Smith County in Texas; the Richton Salt Dome and Cypress Creek Salt Dome in Mississippi; and Louisiana's Vacherie Salt Dome.

Among specialists as well as lay-people, opinions remain deeply divided about whether or when a solution to the disposal of high-level wastes and spent fuel will be found. Dr. Helen Caldicott, an influential anti-nuclear activist who is the head of Physicians for Social Responsibility, believes a real solution never will be found: "Even the most brilliant scientist may honestly think he's developed a container that will remain intact for a half million years. . . . [B]ut he's going to be dead in 30 years; he will never live to verify his hypothesis," she said at a DOE public hearing on nuclear waste management in 1978.

On the other hand, an authority on nuclear waste who once was connected with the Natural Resources Defense Council — a frequent critic of the industry — has offered the opinion that if researchers are willing to put in the effort, they should be able to develop an adequate repository. A scientist has claimed that prospective repositories are going to be "the most over-engineered installations ever built in the history of humanity" and that "there is absolutely no reason why we shall not be able to engineer safe repositories in many places."

Worldwide Safety Problems

The problems afflicting the nuclear industry are by no means unique to the United States. Canada once claimed that its CANDU heavy-water reactors were virtually immune to sudden accident, but in the summer and fall of 1983 an unprecedented series of large leaks occurred at several reactors, including one leak where radioactive water was spilled into Lake Ontario. The leaks forced costly shutdowns and undermined Canadian efforts to promote sales abroad of the CANDU as an alternative to the light-water reactors manufactured in the United States, Western Europe, and Russia.

Nuclear utilities in Western Europe appear to have a relatively good safety record, and industry representatives often claim that they have learned to build U.S.-model reactors better than they are made in the United States. But in principle any one of the reactors that now dot the major European rivers and coasts could experience a TMI-type accident. In countries such as France, where the central government's powers are extensive and access to official information is limited, one cannot help but wonder whether the industry's record is as unblemished as officials say. If a fundamental defect were to turn up at any of France's nuclear power plants, which are being built in large numbers according to a standardized design, the effects would be far-reaching.

Russian officials used to claim that nuclear safety problems were unique to capitalist countries where — they say — a greed for profits tempts businesses to cut corners. In the summer of 1983, however, a political battle over safety and engineering standards broke into the Moscow press when the ruling Politburo reprimanded nuclear officials for "flagrant violations of state discipline." Apparently, Soviet officials were unhappy about mishaps in the construction of "Atommash," a giant industrial facility on the Don River, which was to be the country's principal reactor manufacturer. After rebuking nuclear officials, the Politburo ordered the establishment of a new state agency, similar to the United State's NRC, to monitor and regulate the Soviet nuclear industry.

Until recently, Russian reactors were built without containment structures, and when they were exported to countries such as Finland local authorities bought containments from western suppliers. As for the Soviet reactors located in Russia itself, there is no way of knowing whether they have operated as flawlessly as the government says, especially because accidents of all kinds usually are not reported in the Soviet press. The government still has not explained how a large tract of land in the Urals near Sverdlovsk came to be devastated in 1957. Zhores A. Medvedev, a leading Russian scientist and science writer, charged in a book published four years ago in English that the devastation resulted from a chemical explosion at a nuclear waste repository.[7] Zhores Medvedev is a political dissident who now lives in London, and when he first made his allegations he was ridiculed by British nuclear officials.

In countries such as West Germany and Sweden, authorities have been unable to proceed as rapidly as they expected with plans for nuclear waste repositories — partly because of political opposition and partly because of technical problems similar to those encountered in the United States. Governments in many of the advanced industrial countries remain committed to the idea that reprocessing is an essential aspect of the waste disposal process, and a number of them have sent spent fuel to factories in England and France for treatment. But the plants operated by United Reprocessors GmbH at Windscale and La Hague have suffered frequent mishaps, and critics connected with labor unions have claimed that they pose unacceptable dangers to the health of their workers.

In the United States opinion is closely divided about whether or how much reprocessing of spent fuels might facilitate the spread of nuclear weapons. Many specialists in the field have argued that separation of plutonium from the highly radioactive spent fuel would make it easier for terrorists to get the crucial material needed for bombs and to tempt governments, which otherwise might have remained nuclear-free, to go nuclear.

Some people regard the risks connected with spent fuel, waste disposal, and nuclear proliferation as so very great that they consider the debate over nuclear energy to be, ultimately, a moral issue. Better to have no nuclear energy at all, they say, than to tempt unruly groups and untrustworthy governments with the materials needed for bombs. Or better, at least, not to reprocess fuels anywhere in the world.

Since the mid-1970s these issues have been hotly debated in the United States government, universities, and policy think-tanks. Meanwhile, at the grass-roots level concern about the spread of nuclear weapons has begun to grow to the extent that presidential candidates now consider it worthwhile to emphasize their positions on proliferation. Nearly every candidate maintains that his policies will be more effective than any other's, but it is not easy to evaluate such claims because nuclear proliferation involves a formidable array of political, legal, and technological complexities.

Notes

1. Atomic Industrial Forum, "Low-Level Radiation," April 1983.
2. Josef Rotblat, "The Risks for Radiation Workers," *The Bulletin of the Atomic Scientists* (December 1977): 31-36.
3. Ibid., 54-56.
4. Bernard L. Cohen, "What Is the Misunderstanding All About?" *The Bulletin of the Atomic Scientists* (February 1979): 55.
5. Karl Z. Morgan, "Cancer and Low-Level Ionizing Radiation," *The Bulletin of the Atomic Scientists* (February 1979): 58.
6. Peter Metzger, *The Atomic Establishment* (New York: Simon & Schuster, 1972), Chap. 4.
7. Zhores A. Medvedev, *Nuclear Disaster in the Urals* (New York: W. W. Norton, 1979).

Part II

Nuclear Proliferation

British Prime Minister Winston Churchill, U.S. President Franklin D. Roosevelt, and Soviet Premier Joseph Stalin met at Yalta in 1945 to discuss the future of the post-war world. Together with French President Charles de Gaulle and China's Chairman Mao Tse-tung, they represented the countries that by 1964 would have developed an atomic bomb.

Chapter 6

ORIGINS OF THE NUCLEAR CLUB

The United States government was determined to prevent the spread of nuclear weapons almost from the moment American leaders first began to think about the implications of the atomic bomb. But the prevention of nuclear proliferation proved to be an elusive goal, and, especially during the first two decades of the nuclear era, measures adopted by U.S. governments often turned out to be only partially effective, ineffective, or even counterproductive. Independent critics sometimes accused U.S. leaders of being half-hearted or insincere; the typical response from policy makers usually was to point out that other foreign policy goals often competed for their attention and sometimes took precedence over non-proliferation efforts.

In truth, even when people acted independently and with great single-mindedness to stop the spread of atomic weapons, the results sometimes were quite different from what anybody would have expected. In 1939 when the Hungarian émigré physicist Leo Szilard was beginning to get obsessed with the notion that Nazi Germany might build an atomic bomb, he launched a campaign to get western scientists to stop publishing their discoveries in the field of atomic physics. He ran into resistance among the French physicists who regarded themselves — with good reason — as being in the forefront of the revolutionary new field. But by mid-1940 France was defeated, and, largely as a result of Szilard's efforts, the American and British journals had stopped printing articles on atomic physics.

Two years later, we now know, an alert young physicist in the Soviet Union noticed that articles on atomic science had disappeared from U.S. publications.[1] Putting two and two together, he concluded that the United States had begun a secret project to build an atomic bomb. He wrote to Joseph Stalin, informing the Russian dictator of his conclusions, and by the end of 1944 the Soviet Union had 100 atomic physicists at work in a laboratory headed by Igor Kurchatov. Meanwhile, Stalin was receiving intelligence reports about the Manhattan Project.

When the war ended Americans learned that Germany was not nearly as advanced in its atomic research as people such as Szilard had feared. But Russia, which was fast emerging as the main antagonist of the United States in world affairs, was poised to take off on a crash effort to build its own atomic bomb.

Soviet and British Bombs

During World War II the Danish physicist Niels Bohr urged President Franklin D. Roosevelt and British Prime Minister Winston Churchill to inform Stalin of the bomb and to seek an agreement on the international control of atomic energy. Bohr is universally considered one of the greatest physicists of the 20th century, but he was not eloquent in personal conversation, and the two Allied leaders were not impressed by his arguments. Roosevelt and Churchill reached a secret agreement during the war to consult on atomic policy. They did not make any binding decisions about whether or how the bomb would be used — if it worked — but they plainly did not intend to give Stalin a voice in atomic policy.

When President Harry S Truman, after Roosevelt's death and the defeat of Germany, met with Churchill and Stalin at Potsdam, Germany, to discuss postwar international arrangements, word arrived that the bomb had been successfully tested. Truman, after consultations with Churchill and Secretary of State James F. Byrnes, privately informed Stalin that the United States had developed a new weapon "of unusual destructive force." Stalin replied that he hoped the United States "would make good use of it."

According to Marshal Georgi A. Zhukov, commander of the Soviet forces in World War II, Stalin told his aides after the meeting with Truman that it would be necessary to speed up Kurchatov's atomic research right away.[2] In mid-August after the bombing of Hiroshima and Nagasaki, Stalin held a meeting with the managers of the Soviet munitions program and with Kurchatov, director of the atomic laboratory. Stalin is reported to have said: "A single demand of you, comrades: Provide us with atomic weapons in the shortest possible time. You know that Hiroshima has shaken the whole world. The equilibrium has been destroyed. Provide the bomb — it will remove a great danger from us."[3]

In the United States and to a lesser degree in England there were lingering hopes that an agreement on international control of atomic energy could be negotiated so that an atomic arms race with the Soviet Union would be prevented. By early 1946 the big powers had agreed to open a United Nations inquiry on international control, and the U.S. government had established committees to investigate atomic issues. An advisory committee headed by David Lilienthal, former head of the Tennessee Valley Authority and future head of the Atomic Energy Commission (AEC), produced a plan that in turn became the basis for a proposal that Bernard M. Baruch presented to the U.N. Atomic Energy Commission on June 14, 1946. Baruch, a financier, had served as adviser to several presidents in the areas of economic policy, war mobilization, and war demobilization.

The Lilienthal-Baruch plan called upon all nations to transfer to an international authority ownership and control of all atomic materials and activities; once the authority was established, nations already possessing atomic weapons (then only the United States) would destroy them. The Soviet government immediately rejected the Lilienthal-Baruch plan because it required the Soviet Union to give up independent control over its atomic research before the United States had discarded its arsenal of nuclear weapons. Andrei Gromyko, the Soviet ambassador to the United Nations in 1946, presented an alternative proposal that would have required the United States to abolish its nuclear arsenal before an international authority was estabished. This proposal was equally unacceptable to the U.S. government, and negotiations quickly came to an impasse.

British leaders took no independent initiatives to break the deadlock. They were beginning to worry about potential restrictions on their own atomic research, and they were deeply skeptical about the real possibility of achieving an agreement. In the fall of 1945 an official committee advised the British government that it probably would not be possible to reach an agreement that restricted the freedom of any of the major powers to produce atomic weapons. By major powers the committee meant the five leading allies that had won World War II — the Soviet Union, Britain, the United States, France, and China.[4]

On Aug. 29, 1949, the Soviet Union tested its first nuclear explosive. Despite Congress's imposition of tight secrecy on U.S. atomic research, and despite Truman's belief that the United States would be able to maintain a nuclear monopoly for 10-20 years, it had taken the Soviet

97

U.S. and Soviet A-Bomb Programs

January 6, 1939. Discovery of nuclear fission by Otto Hahn and Fritz Strassmann in Berlin, capital of Hitler's "Third Reich."

January 1939. Niels Bohr informs scientists at Princeton, N.J., of the breakthrough in fission.

August 1939. Albert Einstein writes to Franklin D. Roosevelt about the possibility of building an A-Bomb.

April 1940. U.S. scientists agree not to publish papers that might help Germany develop an A-bomb.

June 1940. Roosevelt names Vannevar Bush to head National Defense Research Committee and gives it jurisdiction over A-bomb work.

December 6, 1941. Bush recommends launching a crash program to build an A-bomb.

August 1942. "Manhattan Project" established. J. Robert Oppenheimer named scientific director; Gen. Leslie R. Groves made the management chief.

July 16, 1945. First A-bomb test.

August 6 and 9, 1945. Atomic bombings of Hiroshima and Nagasaki.

February 1939. Soviet scientists learn of the discovery of fission from foreign science journals.

Late 1940/Early 1941. N. N. Semenev, a Russian scientist, writes to Soviet government warning of A-bomb.

1942. Soviet government names S. V. Kaftanov "plenipotentiary" for science and sets up Scientific-Technical Council to assist him.

Fall 1942. Scientist G. N. Flyorov writes to Stalin that lack of publications from top U.S. scientists indicates secret secret A-bomb project.

October 1942. Soviet A-bomb project established. Igor Vasil'evich Kurchatov named scientific director; L. P. Beria, the chief of the Soviet secret police, made responsible for overall management.

August 1945. Stalin orders a crash program to build a Soviet A-bomb.

August 29, 1949. First successful Soviet A-bomb test.

Union just four years to develop its bomb from the time Stalin ordered a crash program — almost exactly the length of time it took the Manhattan Project to build the U.S. bomb. No doubt the Russian team got useful information from spies such as Klaus Fuchs, a Manhattan Project scientist who confessed in 1950 to having passed secrets to the Soviet Union. Most historians believe, however, that such information was only of marginal help. It probably enabled the Russians to make some engineering choices more quickly, but it certainly did not tell them anything they could not have figured out on their own.

Historians still debate whether the Soviet Union could have been dissuaded from building the bomb if the U.S. government had been more open during the war and if it had tried harder to negotiate a compromise with Russia in 1945-1946. On the one hand, it seems difficult to believe that Stalin would have agreed to any arrangement that left the United States with the technology to build bombs while ensuring that the Soviet Union would be without it. Even if the U.S. facilities were put under an international authority, Stalin would have worried that a U.S. government could repudiate the agreement and resume at any time with bomb production.

On the other hand, given the U.S. policies, it was all but a foregone conclusion that Russia would go nuclear. As explained by Margaret Gowing, author of several books on England's nuclear program: "If Russia had been formally consulted about the bomb during the war . . . it might have made no difference. The fact that [Russia] was not [consulted] guaranteed that the attempts made just after the war to establish international control . . . were doomed." [5]

All in all, the U.S. policy of trying to maintain a nuclear monopoly after World War II may have affected England's atomic policy more than the Soviet Union's. In November 1945 the leaders of the United States, England, and Canada reached a rather flimsy agreement to continue with the nuclear cooperation established during the Manhattan Project. But by this time members of the U.S. Congress already were drafting legislation that would impose tight restrictions on the disclosure of atomic information to any other countries.* British scientists and policy makers were deeply offended by the McMahon legislation, which reminded them unpleasantly of the isolationist policies Congress had adopted between the two world wars.

---

* See Chapter 2.

Even before the McMahon Act passed Congress, however, the British Labor Party government headed by Clement Atlee already was making rapid strides toward developing an independent nuclear capability. The government decided to proceed with the construction of a plutonium production reactor, the most efficient route to a bomb, in December 1945. It authorized construction of a uranium enrichment plant in fall 1947, and it ordered an atomic bomb to be built in early 1948. These decisions were taken without public debate and with only the briefest possible mention in parliamentary records. As Margaret Gowing has shown, the decision to build the bomb "emerged" from a body of widely shared assumptions. It was not, she said,

> a response to an immediate military threat but rather something fundamentalistic and almost instinctive — a feeling that Britain must possess so climacteric a weapon in order to deter an atomically armed enemy, a feeling that Britain as a great power must acquire all major new weapons, a feeling that atomic weapons were a manifestation of the scientific and technological superiority on which Britain's strength, so deficient if measured in sheer numbers of men, must depend.

Thus when Churchill returned to power in October 1951, he found "with a mixture of admiration, envy and the shock of a good parliamentarian" that his Labor government predecessors had spent nearly 100£ million on atomic facilities without informing Parliament.[6]

The first British A-bomb test was in October 1952. By this time both the United States and Russia had embarked on crash programs to build hydrogen bombs (H-bombs), which were expected to be enormously more powerful than atomic bombs. The first U.S. test of a "superbomb" also was in October 1952, and the first Soviet test of a truly comparable weapon was in November 1955. Britain's Conservative government ordered construction of an H-bomb on March 30, 1955, with the support of the National Executive Committee of the Labor Party. British party leaders believed that possession of a full-fledged nuclear deterrent would enable the country, if necessary, to pursue foreign policies independently of the United States. Still more important, perhaps, was their view that a full-fledged nuclear deterrent would be a "ticket to the table" — a guarantee that Britain would be included in summit conferences where major world issues were resolved.[7]

France and China Join the Club

French atomic scientists, like the British, had contributed greatly to atomic research before the war, and they, too, resented the restrictions imposed by the U.S. Congress after the war. A team headed by Frederic Joliot-Curie was assembling the materials to build a nuclear reactor when Germany invaded France — long before scientists in the United States had begun to construct a reactor at the University of Chicago. In fact, Joliot-Curie's group filed for a patent on the nuclear reactor at the beginning of the war. After the war the French scientists tried without success to win legal recognition for their claim that they had invented the nuclear power plant.[8]

The first postwar governments in France hedged on whether the country should build atomic weapons, but an atomic energy agency — the Commissàriat a l'Energie Atomique (CEA) — promptly was established. Joliot-Curie was named scientific director, but he operated under the supervision of an administrative chief. While the French authorities recognized that Joliot-Curie was one of the most brilliant physicists of the 20th century, they did not altogether trust him because he was a communist and had worked with communists against Hitler in the French Resistance.

During the first years of French atomic research, Joliot-Curie was careful to make assurances of loyalty to the government. But in April 1950, Joliot-Curie got himself into serious trouble by making the following statement at a national convention of the French Communist Party: "Never," he said, "will progressive scientists, never will communist scientists, give a particle of their knowledge for a war against the Soviet Union." [9] According to Lawrence Scheinman, an authority on the history of atomic energy in France, it is not known exactly why Joliot-Curie made this statement. It is worth noting, however, that Europeans were increasingly fearful in 1949-1950 that the U.S. government might launch a nuclear strike against Russia. When the Korean War broke out in 1950, British leaders rushed to Washington to advise Truman against retaliating with an atomic attack on the Soviet Union. What worried the Europeans most was the hysterical anti-Soviet mood that had swept the United States in the late 1940s and early 1950s.

The Russian atom bomb test of 1949, together with disclosures that spies working for the Soviet Union had infiltrated the Manhattan Project, had created a highly explosive situation in the United States. The

U.S. public was completely unprepared for news of the Russian test, and many Americans sought to put the blame on "traitors" in high places. The stage was set for the witch-hunt that Sen. Joseph McCarthy, R-Wis., began to conduct in February 1950, just days after Klaus Fuchs's confession in London. Meanwhile, J. Robert Oppenheimer, as head of the General Advisory Committee to the AEC, was finding himself in political hot water by vigorously opposing the decision to undertake a crash program to develop a U.S. H-bomb. While the Oppenheimer proceedings ended up centering on questions of character, Oppenheimer's pre-war connections with communists contributed to his political difficulties.

Joliot-Curie had never made a secret of his communist ties, but his statement in April 1950 that he would not contribute to the construction of an anti-Soviet A-bomb cost him his job. The French authorities proceeded to reorganize their atomic energy commission, and they made it clear that loyalty to the government would figure prominently in their selection of new personnel. A year later, while the CEA still was being re-organized, the authorities decided to build France's first plutonium production reactor — the most efficient first step to a bomb. During the following years a schedule for plutonium production was included in France's five-year economic plan. Bertrand Goldschmidt, a leading French atomic scientist and historian, said that military motives were "present, and undoubtedly, predominant, in the minds of those who inspired and were responsible for the plan." [10]

By 1955 France had built three plutonium production reactors, and the French military had begun to contribute to funding for the atomic program. Governments continued to hedge on the question of whether a bomb ultimately would be built, but they saw to it that the wherewithal for bomb production was assembled. The effort was especially energetic after a socialist government headed by Guy Mollet took office in 1956. In France, as in England, the foundation for production of atomic weapons was laid by left-wing governments.

In the negotiations that led to the creation of EURATOM, a European atomic development authority founded as an organ of the European Economic Communities in 1957, Mollet was careful to see that France's legal and material ability to manufacture atomic weapons would not be compromised in any way by membership in the organization. According to Lawrence Scheinman, "the movement for atomic European union would have suffered defeat at Parliament's hands if the Mollet

The Oppenheimer Case

J. Robert Oppenheimer, after serving as director of the Manhattan Project and chairman of the Atomic Energy Commission's "General Advisory Committee," was exiled from policy-making circles in 1954 when the AEC relieved him of his security clearance. At the time, he was widely seen as a victim of Sen. Joseph McCarthy's efforts to expose communists in government, but the facts of the matter were much more complicated.

Though a brilliant and inspiring teacher, Oppenheimer also was an eccentric and sometimes arrogant person who antagonized many people, including President Harry S Truman. In a meeting between the two men after the war, Oppenheimer told Truman that he felt as though he had blood on his hands because of the atomic bombing of Hiroshima and Nagasaki. Truman, who was contemptuous of sentimentality in men, asked him whether he would like to have a handkerchief.

Oppenheimer often irritated his colleagues by taking unusual or unexpected positions on issues, but it was his opposition to development of the hydrogen bomb that was the main factor behind his troubles in 1953-1954. After Eisenhower took office, the man he appointed to be new chairman of the AEC was persuaded to launch a security investigation on Oppenheimer. Initially, the investigation centered on ties Oppenheimer had with communists in the 1930s, but these connections already had been thoroughly examined by the government during World War II, and nothing new had been turned up. By the time the hearings ended, they had come to focus on Oppenheimer's character and general trustworthiness. Ironically, it may have been Oppenheimer's decision to inform the government about a personal friend's efforts at espionage that contributed most to the impression that Oppenheimer was not completely honest and trustworthy. Testimony by Edward Teller, the key figure in the development of the U.S. H-bomb, also was damaging to Oppenheimer's case. Teller told the commission, "I feel that I would like to see the vital interests of this country in hands which I understand better and therefore trust more."

Source: Murray Kempton, "The Ambivalence of J. Robert Oppenheimer," *Esquire* (December 1983): 236-248.

government had not conceded to the powerful political forces operating within Parliament by guaranteeing the retention of French rights and capacity to undertake military atomic development.[11]

In November 1956 the French authorities decided that the CEA should undertake preliminary studies for a bomb test, and the final decision to build a bomb was taken in 1958. Gen. Charles de Gaulle informed U.S. Secretary of State John Foster Dulles of the French decision within five months of taking office as France's leader. Up until this point, many of the crucial decisions that had given France the capability to produce atomic bombs had been quietly made by top Cabinet members and atomic energy officials, without public debate or public disclosure.

A number of factors probably contributed to the French government's final decision to "go nuclear" and to make no secret of it: the Suez crisis of 1956, in which England and France were forced to back down from an intervention in Egypt in the face of an open nuclear threat from the Soviet Union and strong opposition from the Eisenhower administration; the Soviet intervention in Hungary, which took place at the same time the crisis in the Middle East was unfolding; and the Russian tests of the world's first inter-continental ballistic missiles in 1957. When the Soviet Union launched Sputnik in October 1957, it became plain that Europe and the United States could soon be targeted by long-range Russian missiles capable of carrying atomic and hydrogen bombs. The rocket that put Sputnik into orbit also could be used to deliver nuclear warheads to far away countries.

Among the many factors that influenced General de Gaulle to pursue highly independent defense policies was the privileged status that the British enjoyed in Washington, not only during the Eisenhower years but in the 1960s as well. Before England went nuclear U.S. relations with the British government had been highly ambivalent in the field of atomic energy. American scientists were keen to gain access to British expertise in certain areas of atomic research, and the U.S. government was eager to take advantage of the Commonwealth's uranium reserves in South Africa and Canada, which were among the largest known at the end of World War II. But the government was not eager to see England go nuclear, and Congress was determined to keep a tight lid on American nuclear know-how.

After England went nuclear, U.S. governments adopted a much more open-handed attitude toward the British. In 1957 the U.S. and British governments reached an agreement providing for the installation in England of 60 intermediate-range missiles under a "dual key" system, which was designed to give the British a veto over their use. In 1958 the U.S. Congress revised atomic energy legislation to facilitate the exchange of nuclear information with England. In 1962 when the United States agreed to supply England with a completely independent submarine-based missile force, General de Gaulle was so incensed that he vetoed British entry into the European Economic Community. By this time France had tested its first atomic bomb and was preparing to develop hydrogen bombs. By 1967 when France disassociated itself from the North Atlantic Treaty Organization, General de Gaulle had established a solid reputation in Washington as the truant of the Atlantic alliance.

In Moscow, Soviet leaders had even more serious problems with Mao Tse-tung, China's revolutionary leader. When the Chinese communists took power in 1949 they were grateful to the Soviet Union for having provided inspiration and support during a long civil war, but they also were determined to restore China's dignity in the world. For a century China had been at the mercy of foreign intruders who allied themselves with feuding warlords to advance their interests.

Relations between China and Russia began to deteriorate sharply in the late 1950s after Soviet leader Nikita Khrushchev denounced Stalin at the 20th Party Congress. The Chinese continued to regard Stalin as a patron saint, and Mao began to think of himself as Stalin's true successor as world communist leader. The Chinese denounced Khrushchev's halting attempts at rapprochement with the United States — which Mao took to calling a "paper tiger"— and they expressed mounting dissatisfaction with the overall level of Soviet economic aid. High on their list of grievances was Khrushchev's refusal to provide them with the equipment they wanted and needed to build atomic bombs.

Sino-Soviet relations went into a deep chill after Khrushchev's friendship tour of the United States in 1959, and the Chinese began to accuse the two superpowers of forming a "condominium" that would dominate the world. According to Adam Ulam, a leading specialist on Soviet affairs at Harvard, the Chinese felt that their suspicions were borne out by a statement that the Russian Foreign Minister Andrei Gromyko made at the 22nd Party Congress in October 1961: "Our country places special importance on the character of the relations between the two giants, the Soviet Union and the United States," Gromyko said. "If our two countries united their efforts . . . who would dare and who would be in a position to threaten peace? Nobody. There is no such power in the world." [12]

In 1964, the year after Russia and the United States negotiated their first significant arms control agreement (the partial test ban treaty), China exploded its first atomic bomb. It tested its first hydrogen bomb in 1967, much sooner than expected. France tested its first superbomb in 1968, eight years after its first atomic bomb. But China, despite its economic backwardness, had developed its bomb in just three years. World leaders were impressed by China's effort, and they began to take the Chinese more seriously. Like the British and French before them, the Chinese regarded the atomic bomb as a "ticket to the table" — and they were not wrong.

With the Chinese nuclear tests, what has come to be known as the "nuclear club" was completed. The United States was not happy to see the Soviet Union go nuclear, and neither of the giants was happy to see England, France, and China join the club. Both tried hard, each in its own camp, to maintain exclusive control over the bomb, and both paid high prices for that policy. But the non-proliferation policies of both superpowers were inconsistent and often counterproductive, and neither was willing or able to keep England, France, and China out of the club. In their hearts, one suspects, the U.S. and Soviet leaders may have believed that all five of World War II's victor powers really deserved to join.

However that may be, it was another matter when atomic policy was applied to the losers of World War II. Nobody wanted to see the bomb in the hands of the Germans, who had demonstrated all too vividly what can happen when a fanatic or irresponsible leader gains control of a massive military machine.

The Germany Problem

After World War II the U.S. occupation authorities imposed a constitution on Japan that barred that country from acquiring a military force capable of launching offensive operations. The Japanese limitations have remained intact to the present day, in part because they are widely supported by the Japanese themselves who were the first, and so far the only, victims of an atomic attack.

When the war ended the victorious powers were almost equally determined to keep Germany weak and demilitarized. But as cold war tensions invaded Europe during the late 1940s, western leaders began to consider the rearmament of West Germany as essential to European security. The plan in the early 1950s was to incorporate the West German army into a joint "European Defense Force." In August 1954, however, the French Parliament rejected the force, and two months later leaders met in London to make alternative arrangements. At the London meetings, as a precondition for West Germany's rearmament, Konrad Adenauer pledged not to build or acquire nuclear weapons.

Adenauer was the U.S. government's choice to lead postwar Germany, and his London pledge was treated with a good deal of solemnity. Apparently by pre-arrangement, then Secretary of State Dulles said that the pledge had been made *rebus sic stantibus*. Dulles's reference was to a basic principle in international law, which holds that an agreement is valid only as long as fundamental conditions relevant to the agreement remain unchanged. Adenauer agreed that the pledge was indeed made with this standard proviso. During the following years he and his colleagues would trade skillfully on fears that Germany might in fact change its mind and go nuclear.

In May 1955 West Germany's non-nuclear pledge was formalized in the London-Paris Accords. From a legal point of view West Germany's promise made it the world's first officially non-nuclear country. In actual fact, however, nuclear weapons already were being deployed on West German soil. The weapons, to be sure, were controlled by the NATO commander, not West Germans. But it is noteworthy just the same that in renouncing nuclear weapons the West Germans did not renounce a nuclear defense.

The decision to deploy U.S. nuclear weapons in Europe was taken at a NATO meeting in 1952. Two years after that, U.S. atomic energy legislation was revised to facilitate deployment of weapons to NATO allies (and also to permit the dissemination of nuclear information in the context of the Atoms for Peace program). Planning for the deployment of nuclear weapons in Germany proceeded apace, and in June 1955 military maneuvers involving the hypothetical use of "tactical nuclear weapons" were held in West Germany. Tactical nuclear weapons, for all practical purposes, were indistinguishable from strategic nuclear weapons when they were first deployed. In theory, they were earmarked for battlefield use, but in a densely populated country such as West Germany, it would be difficult or impossible to save cities from their effects, even if Soviet forces abstained from nuclear retaliation. When tactical nuclear war maneuvers were held in West Germany, the result was a political firestorm.

The June 1955 maneuvers, known as "Carte Blanche" or "blank slate," assumed that 300 nuclear warheads had exploded between Hamburg and Munich — an area about the size of Pennsylvania. In July and December 1955 parliamentary debates about tactical atomic warfare were broadcast to millions of West Germans who heard the Social Democratic Party (SPD) denounce the government's military policies. At that time the SPD's leadership considered West German rearmament unnecessarily provocative, and they thought it doomed any hope for German reunification, which they wanted to promote by means of neutralist policies.

In March 1958 following another debate on tactical nuclear warfare in the Bundestag, the West German parliament, the SPD launched an extra-parliamentary campaign against nuclear weapons. They called the campaign "Kampf dem Atomtod," or battle against the atomic death. The following year, however, a more conservative leadership took over the party and persuaded the SPD to back the policy of rearming West Germany within the framework of the NATO alliance.

Opinions differ among scholars as to whether there was any serious political pressure during this period for West Germany to follow in the footsteps of England and France by acquiring an independent nuclear force. One historian reviewing the years 1954-1966 found no "official or discoverable effort toward acquiring a national [nuclear weapons] production capability or a capacity for hair-trigger development," (in other words, the capability of manufacturing nuclear weapons very quickly should conditions change radically.)[13] On the other hand, another historian warned that "students of the German political scene are well advised not to take publicly stated German views as being a necessarily reliable indication of the opinion that Germans hold in private."[14]

Publicly, Adenauer pledged not to build nuclear weapons, but privately he often hinted that it might be necessary to build them some day. Franz Josef Strauss, who served first as Adenauer's Minister of Atomic Energy and then as Defense Minister, avidly favored the acquisition of nuclear technology and the deployment of tactical nuclear weapons. In 1958 Strauss reportedly held discussions with his French counterpart in which he offered financial assistance for France's atomic research in exchange for assured access to French atomic arms in the event of an emergency.

De Gaulle put an end to such talk, but when the Kennedy administration took office in 1961 Adenauer was an old man and Strauss was his heir apparent. Kennedy officials took the possibility that West Germany might choose to go nuclear very seriously. Thousands of tactical nuclear weapons were being deployed on West German soil under rather ambiguous control agreements, which seemed to give a U.S. general the right to make decisions that could result in the total nuclear devastation of the Federal Republic or its defeat by a foreign power. With the Soviet Union beginning to deploy nuclear-armed long-range missiles, West Germans were beginning to wonder whether U.S. leaders could be counted on to sacrifice American cities to defend Hamburg, Frankfurt, or Munich. West Germans, in short, did not know what to be more afraid of: that the United States would respond to a Soviet attack with nuclear weapons, or that it would not.

As a means of easing German anxieties, Kennedy officials toyed with the idea of setting up a multilateral nuclear-armed naval force in which the West Germans would be included. But the idea promised only an illusion of greater security to West Germany, and it never got off the ground. Moscow experimented with a similar notion, partly in retaliation and partly to constrain Chinese ambitions, but they were equally unsuccessful.

Inevitability of the Nuclear Club

When the superpowers tried to exclude the other countries that won World War II from the nuclear club they failed. To a very great extent their efforts to stop proliferation actually provoked the club's expansion. Classification of nuclear research in World War II tipped off the Soviets to the Manhattan Project, and the policy of excluding Soviet leaders from discussions of the atomic bomb undermined any hope — however feeble — for postwar control efforts. When Congress drafted the McMahon Act, out of a desire to keep nuclear information a U.S. secret, England put its atomic research into high gear. When the Eisenhower administration embraced the British government as a partner in the nuclear club, it helped spur France's nuclear ambitions. Russia's refusal to give China nuclear aid only prompted the Chinese to redouble their atomic efforts.

By the early 1960s all the major powers that won World War II had gone nuclear, and it was beginning to seem inevitable that the losers would soon follow. But at this point the giants decided for the first time to take concerted action. Building on the international safeguards system that grew out of Eisenhower's Atoms for Peace proposal, the superpowers drafted a treaty to prevent the proliferation of nuclear weapons.

Notes

1. See David Holloway, "Entering the Nuclear Arms Race: The Soviet Decision to Build the Atomic Bomb, 1939-1945," The Wilson Center, Working Paper No. 9, Washington, D.C., 1979.
2. See David Holloway, *The Soviet Union and the Arms Race* (New Haven: Yale University Press, 1983), 15.
3. Ibid.
4. See Margaret Gowing, *Independence and Deterrence: Britain and Atomic Energy, 1945-52* (London: Macmillan Publishing Co., 1974), 70-71.
5. Margaret Gowing, "Science and Politics," The 8th J. D. Bernal Lecture, Birkbeck College, London 1977, 11.
6. Gowing, *Independence and Deterrence*, 406.
7. See Andrew J. Pierre, *Nuclear Politics: The British Experience with an Independent Strategic Force* (London: Oxford University Press, 1972), 104-105.
8. See Spencer R. Weart, "France and the Origins of Nuclear Energy," *The Bulletin of the Atomic Scientists* (March 1979), 41-50.
9. Quoted in Lawrence Scheinman, *Atomic Policy in the Fourth Republic* (Princeton, N.J.: Princeton University Press, 1965), 41.
10. Bertrand Goldschmidt, *The Atomic Adventure*, trans. Peter Beer (London: Pergamon Press, 1964).
11. Scheinman, *Atomic Policy*, 163-164.
12. Adam B. Ulam, "Lost Frontier," *The New Republic*, Nov. 21, 1983, 10.
13. Catherine McArdle Kelleher, *Germany and the Politics of Nuclear Weapons* (New York: Columbia University Press, 1975), 277.
14. Hans Speier, *German Rearmament and Atomic War* (Evanston, Ill.: Row, Peterson & Co., 1957), ix.

A Rockwell International chemical technician examines a "button" of plutonium at the U.S. government's Rocky Flats plant near Denver. Such materials are monitored by the International Atomic Energy Agency to safeguard against their use in the manufacture of atomic weapons.

Chapter 7

INTERNATIONAL SAFEGUARDS
AND INDIA'S CHALLENGE

The first serious and constructive step to curb nuclear proliferation was taken in 1953, the year after the United States tested its first hydrogen bomb and England exploded its first A-bomb, completing the inner circle of the nuclear club. Speaking at the United Nations in December 1953, President Dwight D. Eisenhower proposed what he called an "Atoms for Peace" plan. The idea was to promote the peaceful use of atomic energy by encouraging countries to acquire and use nuclear fuels and technology under the auspices of a new U.N. agency. Eisenhower's proposal led to the establishment in 1956-1957 of the International Atomic Energy Agency (IAEA), a unit located in Vienna and affiliated with the United Nations.

From the outset critics of Eisenhower's program wondered how promoting the acquisition of nuclear fuels and equipment was supposed to curb the proliferation of nuclear weapons. Critics suspected, with good reason, that a desire to encourage foreign sales of U.S. nuclear technology contributed substantially to the formulation of Eisenhower's plan. Still, Atoms for Peace did lead to an international agreement providing for an unprecedented system of international inspection of nuclear facilities.

Historically, governments have been extremely loathe to give an outside authority independent police powers. When the establishment of the IAEA was negotiated, many officials, politicians, and industrialists in foreign countries worried that safeguards would make nuclear power more expensive and would inhibit development of what seemed to be the energy technology of the future. Among the countries that did not have nuclear weapons, there naturally was grave suspicion that the governments with such weapons were trying to consolidate a monopoly on atomic arsenals.

International Atomic Energy Agency: . . .

The International Atomic Energy Agency (IAEA), a U.N. agency located in Vienna, Austria, is responsible for managing a system of international safeguards that were established in response to President Eisenhower's Atoms for Peace initiative in 1953. The safeguards consist of records kept by parties to the agreements; automatic devices designed to monitor flows of material and to ensure that such devices are not tampered with; and periodic on-site inspections by IAEA personnel. The spot checks by IAEA inspectors are supposed to ensure that equipment and monitoring devices are intact and that stocks of fissionable material correspond to written records.

The IAEA's safeguard system is based on the premise that the spread of weapons can be discouraged only by controlling the materials needed to build weapons because the knowledge of how to build atomic bombs is widely available and becoming more accessible all the time. The objective of the system is to deter the proliferation of nuclear weapons by ensuring, with a reasonable degree of probability, that any attempt to turn nuclear technology to military use will be detected in timely fashion. "Timely warning" is defined in IAEA documents as a period short enough to detect the disappearance of nuclear materials before a government, terrorist organization, or criminal group has time to build a bomb. The IAEA is not set up, however, to actually prevent misuse of materials or to take enforcement action against violators. This inability was demonstrated after an Israeli air raid on an Iraqi nuclear plant in June 1981.

Growing Cooperation Between Superpowers

In negotiations conducted under the auspices of the United Nations in the 1950s and 1960s, Dr. Homi Bhabha, the head of India's nuclear program, often asked how countries with atomic weapons could expect other nations to accept safeguards when they themselves were unwilling to accept them. Bhabha was fond of observing that the safeguard system seemed intended to "ensure that not the slightest leakage took place from the sides of a vessel while ignoring that the vessel had no bottom." [1] In such negotiations Bhabha often found support among members of the "new nations" — countries in Africa, Asia, and

...Watchdog of Nuclear Materials

One of the most challenging aspects of the IAEA's mission has been to devise safeguards for a wide range of nuclear equipment. Light-water reactors are relatively easy to safeguard, for example, because they are reloaded only at intervals of roughly six months and because they must be fueled with low-enriched uranium, which relatively few countries are able to produce. Heavy-water reactors, such as the Canadian CANDU, on the other hand, are refueled continuously, and they run on natural uranium, which can be fashioned into fuel relatively easily. Such reactors need to be watched closely if the IAEA is to stand a good chance of providing the world with "timely warning" of a diversion. Research reactors, especially the larger ones, also can be difficult to safeguard. Many of them operate on high-enriched uranium, which can be used directly in the manufacture of atomic bombs, and some of them are equipped to irradiate materials — including substances such as U-238, which yields plutonium 239 under neutron bombardment.

Facilities that handle large quantities of "critical material" — highly enriched uranium or plutonium — are by far the most difficult to safeguard because a small proportion of their output would suffice for a bomb or bombs. IAEA documents indicate that such "bulk facilities" should be safeguarded, in principle, by means of permanent on-site inspectors. It remains to be seen, though, whether it will be possible to get non-nuclear countries to accept such intrusive inspection, if and when they begin to acquire reprocessing and enrichment technology.

Latin America — that were throwing off the yoke of colonialism. In these regions, which China's leaders dubbed the "Third World," there was resentment about the notion that the big industrial countries could be counted on to handle atomic bombs responsibly, while the small, poor countries could not.

Initially, India and other non-nuclear countries received diplomatic encouragement from the Soviet Union, which adamantly opposed all forms of international inspection in arms control and disarmament negotiations during the 1950s. By the end of the decade, however, Russia began to modify its position — mainly out of concern that Germany or Japan might go nuclear — but also because they were worried about the

Chinese. As early as 1960 the Soviet Union pressed for the strictest possible safeguards on nuclear exports to Japan, while the U.S. government favored looser controls. Three years later, in a decisive shift, the Soviet government voted for the first time to appropriate money for the safeguards system. This reversal in 1963 left India with little choice but to accept IAEA safeguards on a reactor that U.S. firms were to build at Tarapur, near Bombay.[2]

By this time the idea of negotiating a non-proliferation treaty was beginning to receive serious consideration by an 18-member Disarmament Committee that had met in Geneva. In the committee's discussions, the United States and the Soviet Union increasingly found themselves at odds with representatives of new nations. Deeply suspicious of their former colonizers, representatives of Third World countries were inclined to take the view that "if the major powers were left to themselves to devise an acceptable disarmament scheme, no such scheme would ever be found." [3]

By the mid-1960s, when the United States, the Soviet Union, and England were beginning to arrive at a consensus on proliferation, many new and non-aligned nations were predisposed to agree with India's representatives who argued that a non-proliferation treaty would be meaningless and deceptive after so many countries had acquired atomic bombs. Ten years earlier, in 1956, India's foreign minister heartily had urged the U.N. Disarmament Commission to adopt a non-proliferation treaty. But in May 1965, speaking at an IAEA meeting, India's representative said that what the world now needed was "tangible progress toward disarmament including a comprehensive test ban treaty, a complete freeze on production of nuclear weapons and means of delivery, as well as a substantial reduction of existing stocks." [4] The really big problem, India's diplomats grew fond of saying, was not the "horizontal" proliferation of atomic bombs among states but the "vertical" proliferation of nuclear weapons in the countries that already had them.

Undeterred by such views the superpowers continued to narrow their differences on how to go about preventing the horizontal spread of atomic arms. In 1968 they presented a draft non-proliferation treaty to the United Nations. During the next two years governments in many countries debated whether to sign and ratify the treaty. By March 1970, when the Nonproliferation Treaty took force, less than half of the countries represented in the United Nations had signed it.

The Non-Proliferation Treaty

Representatives of the new nations were offended by what they regarded as a "take-it-or-leave it" attitude on the part of the superpowers in the Geneva talks, and the Non-Proliferation Treaty (NPT) was indeed largely the work of the United States, the Soviet Union, and England. Still, negotiators for the nuclear weapons states made a substantial effort to accommodate the wishes of other countries. While they were not able to please everybody, the NPT in its final form reflected elaborate compromises with the non-nuclear industrial countries led by West Germany and Japan and the non-aligned countries led by India and Mexico.

In essence, the NPT promises full access to peaceful nuclear technology, subject to international safeguards, to countries that promise not to build nuclear weapons. Article I bars parties to the treaty from helping other countries get materials for bombs. Article II prohibits non-nuclear countries from building or acquiring nuclear weapons. Article III requires non-nuclear members to accept safeguards on all nuclear energy equipment in which fissionable material — material suitable for use in atomic bombs — is used. Article III also makes the IAEA responsible for managing the safeguard system.

Among the advanced industrial countries there was strong concern that the states with atomic arms would continue to dominate all aspects of nuclear technology if they alone were allowed to work with certain kinds of equipment. Representatives of the industrial countries wondered whether one could keep abreast of developments in nuclear technology unless one stayed on what seemed to be the military cutting edge of the industry. They also worried that international safeguards would raise the cost of nuclear energy for the non-weapons states and make it more difficult for them to compete economically with the weapons states.

Retrospectively, it seems probable that the industrial countries expected nuclear energy to make a more vital contribution to their economies than it actually has delivered. But when the NPT was negotiated, nuclear energy generally was considered the wave of the future, and negotiators for the nuclear weapons states made considerable efforts to accommodate the concerns of countries that hoped to rely heavily on atomic power. Article IV of the NPT guarantees "the inalienable right of all the parties to the treaty to develop research, production and use of nuclear energy for peaceful purposes without discrimination" and obligates parties to "facilitate . . . the fullest possible exchange of equipment, materials and scientific and technological information for the peaceful uses of nuclear energy."

Article V of the treaty goes even further in that it extends the right to acquire and use peaceful nuclear energy even to "peaceful nuclear explosives" — such as atomic bombs meant for use in mineral extraction or large-scale engineering projects such as canal construction. In the late 1960s both the United States and the Soviet Union experimented with peaceful nuclear explosives, and countries such as South Africa and Argentina expressed interest in doing the same. Article V of the NPT says that "the potential benefits from any peaceful application of nuclear explosions will be made available to non-nuclear-weapon states party to the treaty on a non-discriminatory basis and that the charge to such parties for the explosive device used will be as low as possible and exclude any charge for research and development. . . ." Article V specifies that non-nuclear-weapon states should employ peaceful nuclear explosives only "through an appropriate international body."

When the NPT was being negotiated both the United States and the Soviet Union were exploring the use of nuclear explosives for peaceful purposes. But today the NPT's concern about peaceful nuclear explosives seems strange. It appears to show how inflated expectations for nuclear energy were in the late 1960s, and it also shows how far the NPT's negotiators were prepared to go to accommodate those expectations. In the United States and other countries that have nuclear weapons, critics of the NPT often argue that the negotiators went too far. The treaty offers too much in technology, the critics say, in exchange for fragile pledges from non-nuclear countries not to build atomic bombs.

In countries that do not have nuclear weapons, on the other hand, the standard criticism of the NPT is the opposite of the one heard most often in this country. In the Third World, especially, many people think they are asked to offer too much in terms of power and prestige for flimsy promises from the nuclear weapon states to disarm. Under Article VI of the NPT the nuclear weapons states are to "pursue negotiations in good faith on effective measures relating to cessation of the nuclear arms race at an early date and to nuclear disarmament, and on a treaty on general and complete disarmament under strict and effective international control." Leaders of many Third World countries considered Article VI much too vague when the NPT was drafted, and some of them threatened not to sign the treaty unless the nuclear weapons states were prepared to make more specific provisions to curb the arms race. U.S. negotiators responded that the NPT was not intended to be a "commercial treaty," in which countries simply give up one thing to get another. The treaty was intended, they said, to enhance the security of all countries by discouraging their neighbors from building atomic bombs.

Despite this general stance the United States had to make specific commitments to win support for the NPT among some key countries. To allay suspicions that safeguards would raise the cost of nuclear energy and serve as a cover for industrial espionage, the United States offered to open some of its own nuclear facilities to inspection. While this offer had no military significance, it was meant to indicate good will and confidence that the cost of safeguards could be kept within reasonable bounds. In the negotiations that led to the signature and ratification of the treaty by the Federal Republic, the Germans extracted a long list of assurances from the United States that they would be completely unhampered in the development of peaceful nuclear technology. In effect, Germany was guaranteed the right to acquire all the technology that would be needed for atomic bombs in exchange for a strengthened commitment not to actually build them. Germany also got the U.S. government to reaffirm the terms of West German Chancellor Conrad Adenauer's 1954 London pledge, in which he said that Germany would not produce nuclear weapons as long as there were no fundamental changes in the international system. In testimony before the U.S. Senate Foreign Relations Committee in 1969, the secretary of state said that termination of the NATO treaty would constitute the kind of "extraordinary event" that would justify Germany's withdrawal from the NPT.[5]

The secretary of state was referring to Article X of the NPT, which gives any party the right to withdraw from the treaty, providing it gives three months' notice of its decision, "if it decides that extraordinary events, related to the subject matter of this treaty, have jeopardized the supreme interest of its country." Article X enshrines the basic principle of international law — *rebus sic stantibus* — which holds that treaties are only valid as long as fundamental circumstances on which they depend remain unchanged.

The NPT as a Legal Instrument

The Non-Proliferation Treaty suffers from a number of conspicuous imperfections: its tendency to legitimate and perhaps even encourage the spread of the very technologies it seeks to control; its reliance on an under-funded and under-staffed international safeguards authority, the IAEA, which lacks enforcement powers; its provisions that lend legal credibility to the notion that there is a meaningful distinction between "peaceful nuclear explosives" and atomic bombs; and its withdrawal clause.

Even so, the treaty is the centerpiece of international efforts to curb nuclear proliferation. It is the only treaty that negotiators were able to agree on in the late 1960s when world leaders began to agree that some kind of non-proliferation action was necessary and possible. Even today, few specialists on non-proliferation believe that it would be possible to negotiate a substantially strong treaty if the NPT were opened for revision.

Critics of the NPT sometimes talk as though it were a law handed down from above, as though all countries were bound to observe its provisions regardless of whether or not they have signed the agreement. But experts in international law reject this interpretation. International lawyers argue that the NPT is not equally binding on members and non-members. For those countries that have joined, the pledge not to "go nuclear" is about as binding as a marital vow or any other solemn contract; such vows can be broken, but they generally cannot be set aside or violated without some cost.[6] As for the countries that have not joined the NPT, the existence of the treaty exerts a certain moral influence, but it is not binding on them — and will not be — until they are persuaded to join.

In international law, treaties are considered binding on all peoples, regardless of whether their governments have agreed to observe them, only if they embody basic principles or customs that are deemed valid by all decent peoples everywhere. For example, it has been considered dishonorable and abhorrent, at least since the Middle Ages, for armed soldiers to inflict unnecessary harm on unarmed civilians or prisoners. If such offenses occur, the responsible parties can be called to account in an international tribunal, even if their governments have refused to sign and ratify the various treaties covering war crimes.

The world might be better off, many people believe, if everybody considered the spread of nuclear weapons an absolutely bad and unacceptable thing. But in truth, many people are still inclined to think, on balance, that some nuclear proliferation might be better than none. Nations such as Israel and South Africa, Argentina and Brazil, and India and Pakistan still have not decided whether they would be better off with or without atomic bombs. When the NPT was negotiated, France and China refused to take the position that proliferation was necessarily a bad thing, and even today, neither country has joined the NPT. Influential French defense specialists such as Gen. Pierre Gallois have gone so far as to suggest that proliferation is a positive good. If nuclear weapons help prevent war between the superpowers, such theorists ask, why should they not help prevent wars between other countries?[7]

In the United States most people probably are inclined to agree with Stanley Hoffman, a Harvard professor who argued in 1966 that the spread of nuclear weapons would complicate international diplomacy and make some kind of miscalculation or misstep increasingly likely.[8] But such views are far from universally shared elsewhere in the world. Many people take the position that they will have no choice but to acquire atomic bombs if their neighbors continue to assemble and refine nuclear arsenals.

India's Nuclear Test

In May 1974, just four years after the Non-Proliferation Treaty took force, India tested an atomic bomb. The Indian engineers almost certainly obtained fissionable material for the bomb from a nuclear power plant supplied by Canada, in violation of a bilateral safeguards agreement. In what seemed a transparent attempt at self-justification, the Indian government claimed it was merely testing a "peaceful nuclear device" — an atomic explosive suitable, supposedly, for use in civil engineering.

The western press was quick to point out that peaceful nuclear explosives are indistinguishable from atomic bombs, and political leaders denounced India for undermining non-proliferation efforts. But nobody was in a position to take effective action against the Indian government. The Indians plainly did not intend to yield to any pressure short of war, and they always had said they would not sign a non-proliferation treaty unless the nuclear weapons states took significant steps toward disarmament.

The nuclear weapons states would have been in a stronger position to demand reciprocal moves from India if they had in fact adopted far-reaching disarmament measures. But they had not. France and China steadfastly refused (and refuse) to enter into even the most limited arms control agreements. As for the agreements that the United States, the Soviet Union, and England had reached, they could not by any standard be described as disarmament measures.*

Under the circumstances, the nuclear-armed nations were reduced to complaining about India's semi-fraudulent description of its bomb as a "peaceful nuclear device," but even here the superpowers were on shaky ground. At the time of India's test, negotiators for the United States and the Soviet Union were completing work on a "threshold test ban treaty," which would permit each country to use "peaceful nuclear explosives" of less than 150 kilotons.

Because the nuclear weapons nations lacked a firm legal or moral foundation to take action against the Indian government, India's nuclear suppliers, Canada and the United States, adopted limited sanctions, which were expected to have little impact. Canada suspended nuclear cooperation with New Delhi in 1974 and terminated it altogether two years later. The United States temporarily suspended nuclear fuel shipments to India. There was no serious talk in Washington, Ottawa, or elsewhere of imposing more sweeping sanctions on India; nor was there any serious talk of addressing India's complaints about the basic character of the NPT.

Policy Reappraisal and Reform

While the first reactions to the Indian test were half-hearted and un-convincing, the explosion did spark a fundamental reappraisal of non-proliferation policy in a number of countries. In due course, measures were taken on a fairly broad front to tighten up on nuclear exports so that countries such as India would find it more difficult to get the essential technologies and materials for production of atomic bombs.

In April 1975, countries that supply nuclear equipment in the world market resumed semi-secret talks in London about how to tighten export guidelines. Similar discussions had been held before the Indian test by the so-called Zangger Committee, a group of representatives from advanced industrial countries that was chaired by a Swiss official named Claude Zangger. Such meetings were kept quiet because the suppliers were afraid they would be accused of violating Article IV of the NPT, which guarantees the free exchange of nuclear equipment. More generally, the advanced industrial supplier countries were afraid Third World leaders would accuse them of taking an imperialist approach to

---

* See Chapter 9.

the non-proliferation problem — an approach that imposed policies on Third World countries that suited their own interests without the Third World's consent.

Despite such qualms, communist as well as capitalist countries joined the London group, which came to include 15 states.[9] In January 1978 the group released a set of rules the 15 countries had agreed upon to make it more difficult for customers to misuse equipment and materials for atomic bomb production. The relevant nuclear items were included in a seven-page "trigger list," which the suppliers agreed to monitor and enforce.

Under the *London supplier guidelines*, the 15 parties to the agreement said they would require the governments of customer nations to:

- make formal promises that items on the trigger list would not be used to produce an atomic explosive;
- keep the items under effective physical protection to prevent theft or sabotage;
- not retransfer items to third countries without permission from the supplier;
- and accept IAEA safeguards on all trigger-list items.

In the London talks the Soviet government declared itself willing to adopt even tighter guidelines if all other suppliers agreed, but some of the European countries refused to go further.

The Soviet Union favored universal adoption of *full-scope safeguards*, that is to say, a rule requiring nuclear customers to accept IAEA safeguards on all their nuclear equipment, regardless of whether it was bought abroad or when it was obtained. The presence of full-scope safeguards would make it much more difficult for countries to follow India's example — to use unsafeguarded equipment to build bombs out of materials produced in safeguarded facilities. Insistence on full-scope safeguards also would remove an important incentive countries have to join the NPT, in that states outside the treaty only have to accept safeguards on individual imported items, while parties to the treaty have to agree to accept comprehensive safeguards.

As of February 1984 only three countries had adopted a full-scope safeguards rule for nuclear exports: Canada, which led the way in 1976, Australia, and the United States. The U.S. rule requiring customers to accept comprehensive safeguards was incorporated into the Nuclear Non-Proliferation Act of 1978, a complicated bill that underwent several drafts by the time Congress finally passed it. The Non-Proliferation Act directed the executive branch to renegotiate all nuclear cooperation agreements with foreign countries and to make continued cooperation contingent on their acceptance, within 18 months of the bill's enactment, of full-scope safeguards. The bill established the principle that foreign countries should not reprocess, enrich, or re-export nuclear materials obtained from the United States without prior permission of the U.S. government, and it gave the executive branch 24 months to renegotiate nuclear cooperation agreements to include this provision.

In a carrot-and-stick approach, the Nuclear Non-Proliferation Act also contained provisions to encourage the establishment of an international nuclear fuel bank, which would guarantee adequate supplies of fissionable materials for reactors around the world. The point of these provisions was to minimize the economic incentives countries otherwise might have to build reprocessing and enrichment plants, the key facilities needed to supply fissionable materials for bombs. The idea of establishing an international fuel bank was not controversial when the law passed, but not much has come of it.

The most controversial aspect of the Non-Proliferation Act, both in the United States and abroad, was its unilateral character. The act imposed rules on international nuclear commerce, which were not universally accepted as valid or necessary, and it required customers of the United States to agree to the rules whether they liked them or not. The bill gave the Nuclear Regulatory Commission, a body completely independent of the foreign-policy apparatus, the authority to terminate nuclear exports to countries that refused to agree to the new rules within the stated time limits. While the bill gave the president the authority to waive the legislative requirements and allow continued exports, Congress retained the right to override the president's decision by a vote of both houses.

Backers of the Non-Proliferation Act say, in defense of the unilateralist approach, that it was necessary to impose inconvenient rules on the executive branch to ensure that the president would continue to take action on proliferation regardless of who he turned out to be and how he personally felt about the issue. When congressional staffers first began to draft the bill, one of their main objectives was to prod President Gerald R. Ford into acting more aggressively on proliferation.[10] After Jimmy Carter became president, the authors of the act were much more satisfied with the executive branch's non-proliferation efforts, but they decided to retain strict unilateralist language in the bill, partly as insurance against presidents to come.

During the 1976 campaign Carter promised if elected to take a number of specific steps to curb proliferation. In April 1977, less than three months after taking office, Carter announced that his administration would move unilaterally to discourage the development of *plutonium technologies*. The rationale for his policy was that if many countries acquired reprocessing facilities, in which pure plutonium could be extracted from spent reactor fuel, they would be tempted to use the plutonium to construct atomic bombs. To demonstrate that a viable nuclear economy does not depend on reprocessing of spent fuels and recycling of plutonium, Carter said he would terminate federal support for the only reprocessing facility in the United States that was close to starting commercial operations — a plant at Barnwell, S.C. Carter said he hoped his policy would "set a standard" for the world. Later the same month, Carter announced that he would try to get Congress to terminate work on the Clinch River breeder reactor, which was designed to operate on plutonium extracted from spent fuel.

In October 1977 the Carter administration inaugurated an International Nuclear Fuel Cycle Evaluation program, in which some 500 experts from 46 countries participated. The administration's objective was to convince other countries that they could afford to rely on the "once-through" cycle, in which spent fuel from conventional reactors is disposed of without reprocessing, and that they could forego development and deployment of breeder reactors. The administration also hoped to stimulate interest in alternative fuel cycles based on thorium, which promised to be more proliferation-resistant than the plutonium cycle.[11]

During the year leading up to the International Fuel Cycle Evaluation, the search for alternatives to the "plutonium economy" was endorsed by two high-level policy groups: England's Royal Commission on Environmental Pollution, which was headed by Sir Brian Flowers; and the Ford Foundation's Nuclear Policy Study Group, which was chaired by Spurgeon Keeney. The Flowers Commission reported in September 1976 that England "should not rely for energy supply on a process that produces such a hazardous substance as plutonium unless there is no reasonable alternative." The next year the Ford Foundation group concluded that "the United States should work to reduce the cost and improve the availability of alternatives to reprocessing worldwide and seek to restrain separation and use of plutonium." [12]

Impact of Carter Policies

Americans concerned about proliferation had high hopes for Carter, but he did not deliver on some of his 1976 campaign promises, and his policies were less effective than hoped. His efforts to get India to accept comprehensive safeguards were unavailing, and he had little or no success in getting other governments to adopt the U.S. plutonium policy.

Carter's attempt to persuade other governments to forego development of breeder reactors and reprocessing facilities met with almost universal suspicion and sometimes outright hostility. In the Third World people were predisposed to believe that the United States wanted to maintain a monopoly on advanced nuclear technology and to reduce Third World countries to perpetual backwardness. In Europe the prevailing suspicion was that Carter's policies were nothing but sour grapes. The French were strongly inclined to believe that Carter was trying to get them to stop breeder technology because the United States had fallen behind in its effort to develop a commercial fast reactor.[13]

When the International Fuel Cycle Evaluation completed its study in February 1980 it did not endorse Carter's preference for the once-through cycle or recommend other non-plutonium cycles. In retrospect, the exercise may have been more useful than it seemed at the time; subsequent developments have tended to vindicate the Carter administration's viewpoint. Breeder reactors have turned out to be more difficult and expensive to build than Europeans believed in the late 1970s, and the need for them has turned out to be much less urgent than the Europeans claimed. Still, when the fuel cycle evaluation ended it seemed to have been successful only in "avoiding an escalation of tensions" caused by Carter's initiative, as one French writer put it.[14] In other words, it gave the United States the means to present its position without forcing other countries to change their policies. For example, the Carter administration was unable to dissuade Japan from opening a medium-sized reprocessing plant at Tokai Murai, which would produce pure plutonium, and it was unable to block the international shipment of spent fuel to England and France for reprocessing.

In negotiations with India over safeguards and fuel shipments, Carter was less able to retreat without loss of face. In 1977, the year Carter became president, India's prime minister, Indira Gandhi, whose government had tested a bomb in May 1974, was voted out of office. In late 1977 or early 1978, the successor government headed by Morarji R.

Desai proposed a compromise solution to the safeguards dispute: India would accept full-scope safeguards, as required by the Nuclear Non-Proliferation Act that Congress was about to pass, if the nuclear weapons states agreed to 1) a freeze on the production of fissionable material for atomic bombs, 2) a comprehensive nuclear test ban, and 3) target dates for reducing their stockpiles of nuclear weapons. This proposal corresponded closely to the Indian representatives' position in IAEA negotiations during the mid-1960s.

Carter conducted a cordial personal correspondence with Desai, and for a moment it seemed that he might seriously entertain Desai's proposal. The U.S. Arms Control and Disarmament Agency supported a ban on production of fissionable materials, and Secretary of State Cyrus R. Vance wrote a letter to the White House asking Carter to assemble a special team to expedite consideration of the idea. Apparently, Vance hoped to win Carter's approval of the idea quickly so that the United States could announce its decision at the first United Nations Special Session on Disarmament, which was to convene in New York in May 1978.

The Vance letter angered Pentagon officials, however, and news of its contents was leaked to *The New York Times* on April 3, 1978. Defense Secretary Harold Brown and the Joint Chiefs of Staff argued that a ban on production of fissionable materials would not be in the United State's interest because the United States no longer had a commanding lead in the development and production of nuclear warheads. The editorial department of the *Times*, noting that five presidents before Carter had proposed a ban on the production of fissionable materials to the Soviet Union, advised Carter to seriously consider a ban and to prepare a draft proposal for the U.N. conference. Carter decided, however, to follow the advice of his defense secretary and the Joint Chiefs.

Reverting to a policy based on the threat of sanctions, the Carter administration tried for two more years to muscle India into accepting safeguards. After Indira Gandhi returned to power in the summer of 1979, New Delhi took an increasingly intransigent position. The Indian government said that if the United States terminated fuel shipments, Delhi would regard the move as a breach of bilateral agreements and terminate all safeguards on the reactor that U.S. firms had built in Tarapur.

In 1980 Carter waived the requirements of the Nuclear Non-Proliferation Act and permitted continued shipments of nuclear fuel to India. The House voted to override the waiver and suspend shipments, but the Senate upheld the president by a vote of 48-46. Both parties split fairly evenly, but the Democrats voted with the president by a margin of 31-17, while Republicans supported him by a much narrower margin of 24-22. The Republicans, in other words, were more inclined than the Democrats to take a tough line with India, regardless of the consequences.

After President Ronald Reagan took office in January 1981 the dispute with India simmered on. It ended finally with a controversial face-saving compromise in the summer of 1983 when the Reagan administration arranged for France to supply fuel for the U.S.-built reactor at Tarapur.

The Test Ban Disappointment

To supporters of strong non-proliferation efforts, the greatest disappointment of the Carter administration, by far, was the president's decision not to finalize a comprehensive test ban treaty with the Soviet Union, despite the Soviet government's stated willingness to conclude such an agreement and despite Carter's 1976 campaign promise to seek a total prohibition on nuclear tests. Speaking at a U.N. conference in New York on May 13, 1976, Carter said that "the United States and the Soviet Union should conclude an agreement prohibiting all nuclear explosions for a period of five years." Carter referred to the threshold test ban treaty, which the superpowers had signed but not ratified, as a "wholly inadequate step beyond the limited test ban." [15]

Less than a year after Carter took office, on Nov. 2, 1977, Soviet leader Leonid Brezhnev said the Soviet Union was prepared to go beyond the threshold treaty and negotiate a comprehensive test ban. By the summer of 1979 negotiations virtually were complete, but the treaty was running into strong opposition among the Joint Chiefs of Staff and among the engineers at the Los Alamos and Livermore laboratories where all U.S. nuclear weapons are designed. Meanwhile, it had become apparent that the administration was going to have a difficult time getting the Senate to ratify the SALT II treaty, which the United States and the Soviet Union had considered earlier in the year. Carter decided that he could not possibly get the Senate to ratify both SALT II and a test ban, and so he deferred action on the latter. Yet Congress was not convinced that SALT II would prevent a substantial Soviet buildup, and in the end both the treaty and the test ban were defeated.[16]

View of Proliferation Dangers Revised

When the United States and the Soviet Union drafted the Non-Proliferation Treaty in 1967-1968 their main concern was that the Federal Republic of West Germany might soon go nuclear. No doubt the same fear prompted many European countries to promptly sign the treaty. In hindsight, it seems probable that the danger of West Germany going nuclear was always much less than neighboring peoples supposed. The Federal Republic has had no compelling military reason to build nuclear weapons, and Germans were well aware that acquisition of such weapons would make their neighbors even more suspicious of them.

If the danger of Germany going nuclear was overestimated, the danger that other countries might acquire nuclear weapons was perhaps underestimated. The Indians, an acutely status-conscious people who pride themselves on their ancient and sophisticated civilization, found it difficult to see why they were unworthy of being entrusted with atomic bombs merely because they happened to be poor. "It is the nuclear Brahmins that advocate the maintenance of the purity of their caste at the expense of the lower breed," India's representative to the United Nations said in 1977. "It is they who preach that the other states should be required to place all their nuclear facilities and reactors under strict international safeguards, and should not be allowed to conduct PNEs [Peaceful Nuclear Explosives], while they themselves remain free to manufacture nuclear weapons, conduct tests and develop reactor technologies. . . . The notion that some states are inherently more responsible than others is totally rejected by India." [17]

To the East, the Indians have faced a nuclear-armed China since 1964; to the North is nuclear-armed Russia; and in the Indian Ocean to their South, U.S. nuclear submarines armed with nuclear-tipped ballistic missiles are on constant patrol. Influential Indian defense specialists argue that New Delhi can ill afford, under the circumstances, to be without a full-fledged nuclear arsenal. In addition many Indians — like the British, French, and Chinese before them — are inclined to regard a nuclear arsenal as a "ticket to the table."

In 1978 the U.S. government flirted briefly with the idea of admitting India to the table. But it was an idea that threatened to open a can of worms: If the United States gave in to India's demands to prevent the country from proceeding with the construction of an atomic arsenal, what was to prevent other non-nuclear countries from coming to

Washington with their demands? Many foreign policy specialists regarded the idea of trying to deal with India as wildly misguided, and in the end Carter followed their advice. As a result, India remains on the sidelines in the talks affecting arms control and disarmament, and it continues to wonder whether it will become the sixth member of the nuclear club.

Notes

1. See Alvin Z. Rubinstein, *The Soviets in International Organizations* (Princeton, N.J.: Princeton University Press, 1964), 221.
2. For the argument that proliferation cannot be stopped, only managed, see Lewis A. Dunn, *Controlling the Bomb, Nuclear Proliferation in the 1980s* (New Haven: Yale University Press, 1982).
3. Ibid., 220, 224-225, 250. See also Bernard G. Bechhoefer, *Postwar Negotiations for Arms Control* (New York: Columbia University Press, 1961), 358-359.
4. See David A. Kay, *The New Nations in the United Nations* (Baltimore: The Johns Hopkins University Press, 1970), 124.
5. See George Quester, *The Politics of Nuclear Proliferation* (Baltimore: The Johns Hopkins University Press, 1973), 176.
6. See Abram Chayes, "An Inquiry into the Workings of Arms Control Agreements," *Harvard Law Review* (March 1972): 920, 968-969.
7. See Pierre Gallois, *The Balance of Terror* (Boston: Houghton Mifflin Co., 1961), and Kenneth N. Waltz, "What Will the Spread of Nuclear Weapons Do to the World?" in *International Political Effects of the Spread of Nuclear Weapons*, ed. John Kerry King (Washington, D.C.: Government Printing Office, 1979), 165-197.
8. Stanley Hoffman, "Nuclear Proliferation and World Politics," in *A World of Nuclear Powers?*, ed. Alastair Buchan (Englewood Cliffs, N.J.: Prentice-Hall, 1966), 89-90.
9. The original seven members were the United States, the Soviet Union, England, France, Japan, West Germany, and Canada. They were joined by Poland, East Germany, Czechoslovakia, Belgium, The Netherlands, Italy, Sweden, and Switzerland.
10. The Hill staffers who were most active in drafting the bill were Leonard Weiss, who works for Sen. John Glenn, D-Ohio; Connie Evans, an aide to Sen. Charles Percy, R-Ill.; Gerry Warburg, who worked for Rep. Jonathan Bingham, D-N.Y., at the time and now works for Sen. Alan Cranston, D-

Calif.; and Paul Leventhal, who was an aide to former Sen. Abraham Ribicoff, D-Conn.

11. See Harold A. Feiveson, Theodore B. Taylor, Frank von Hippel, and Robert H. Williams, "The Plutonium Economy: Why We Should Wait and Why We Can Wait," *The Bulletin of the Atomic Scientists*, (December 1976): 10-21.

12. See Royal Commission on Environmental Pollution, *Nuclear Power and the Environment* (London: Her Majesty's Stationery Office, 1976) and Nuclear Energy Policy Study Group, *Nuclear Power: Issues and Choices* (Cambridge, Mass: Ballinger Publishing Co., 1977).

13. See William Sweet, "The U.S. Plutonium Policy: Problems and Prospects," *Current Research on Peace and Violence* (1978, no. 3-4), 136-137.

14. See Pierre Lellouche, "International Nuclear Politics," *Foreign Affairs*, (Winter 1979-80): 338.

15. Jimmy Carter, "Three Steps toward Nuclear Responsibility," *The Bulletin of the Atomic Scientists* (October 1976): 8-14.

16. See Jo Pomerance, "The CTB at Last?" *The Bulletin of the Atomic Scientists* (September 1979): 9-10 and William Kincaide, "Banning Nuclear Tests: Cold Feet in the Carter Administration," *The Bulletin of the Atomic Scientists* (November 1978): 48-49.

17. Quoted in Ann Florini, "Nuclear Proliferation: A Citizen's Guide to Policy Choices," (New York: United Nations Association of the USA, 1983), 23.

An F-16 aircraft in vertical climb. Israel used a squadron of American-made F-16s to bomb an Iraqi nuclear power plant on June 19, 1981.

Chapter 8

STOPPING VS. MANAGING PROLIFERATION

To people trained in the school of *Realpolitik*, or power politics, it always has seemed difficult to believe that the spread of nuclear weapons could be confined to just five or six countries. History seems to show, they observe, that states always will acquire the most powerful weapons available. Why should it be different with nuclear weapons? "The genie of nuclear knowledge is out of the bottle," people who regard themselves as realists often point out; sooner or later, they are inclined to believe, every country that amounts to anything will avail itself of nuclear knowledge to build atomic bombs.

To try to stop proliferation altogether, according to this chain of thought, is to fight the inevitable course of history. Efforts to impose universal rules on a hodgepodge of countries with different interests are bound to fail; attempts to stop countries from getting nuclear technology will very likely prompt them to try even harder to acquire materials and equipment; and unilateral restrictions on exports amount to self-inflicted wounds.[1] It would be better, those trained in power politics tend to believe, to yield somewhat to the tide of history. Rather than fight proliferation to the last barricade, they think, we should concentrate on "managing" the spread of nuclear weapons so that U.S. interests are not adversely affected at any given time.

If, by this reasoning, nuclear proliferation is held to a slow enough rate, each new member of the "club" can be initiated into its rites and have time to become acquainted with its rules. Thus the advocates of this approach hope to avoid the missteps and miscalculations that Professor Stanley Hoffman of Harvard University expected nuclear proliferation to bring. In a 1966 article published in *A World of Nuclear Powers?*, Hoffman argued that

> to equate proliferation with peace . . , one would have to assume that
> each nation would find automatically in its nuclear kit not only the

kind of wisdom and responsibility that rule out "irrational" behavior, but the subtle skills that would guarantee success to wise and responsible calculations in a world of uncertainty: a breathtaking assumption, since it means that proliferation would entail an assurance against error as well as against folly.[2]

Some people argue that while wisdom, responsibility, and skill are not to be found automatically in each country's nuclear kit, countries can acquire these qualities, given enough time and care.

When President Richard Nixon took office, skepticism about stopping proliferation reportedly was so profound that the president is said to have ordered officials to make no efforts to persuade other countries to join the Nuclear Non-Proliferation Treaty (NPT). His national security adviser, Henry Kissinger, according to a disenchanted aide, "was convinced that most of the major powers would eventually obtain nuclear weapons and the United States could benefit more by helping them in such efforts than by participating in an exercise in morality."[3]

Critique of Carter Policies

By the time President Jimmy Carter had gone down to overwhelming defeat in the 1980 election, many of his policies had fallen into disrepute, and his approach to non-proliferation was suffering from association with some of his administration's setbacks. According to a commonly heard critique, his non-proliferation policies had disrupted relations with both nuclear and non-nuclear countries, tearing holes in the delicate network of formal and informal agreements and understandings that help restrain the spread of weapons.

"Since 1954," MIT professors Ted Greenwood and Robert Haffa wrote in the spring 1981 issue of *Foreign Policy* magazine, "an international regime for regulating civilian nuclear activities and separating them from military applications has evolved through consensus. . . . Several countries, including the United States, adopted measures intended to strengthen the regime. They did so without international consultation and despite widespread opposition. The result was a disruption of nuclear trade and an erosion of confidence in the efficient and secure functioning of international markets. . . . Stripped of its necessary consensus, the regime will gradually lose its usefulness in helping inhibit weapons proliferation."

Even among people who worked on non-proliferation for Carter, his policies and the policies urged on him by Congress were seen as flawed and too one-sided. Gerard Smith, who served as ambassador at large for non-proliferation negotiations under Carter, and George Rathjens, who worked as Smith's assistant, criticized the administration's policies in a *Foreign Affairs* article in the spring of 1981. They agreed with Greenwood and Haffa that the policies of the previous four years had put too much emphasis "on the unilateral denial of nuclear materials in the form of leverage to prevent proliferation." They thought that such policies had been "based on an inflated view of U.S. leverage" and that "national legislation prescribing universally applicable rules [was] bound to cause trouble."

Smith and Rathjens went on to argue that future U.S. policy should be much more attentive to "the motives that lead nations to want to have nuclear weapons." This could be done, they thought, both by strengthening alliances with countries such as South Korea that otherwise might be tempted to go nuclear and by working more energetically for nuclear arms control between the superpowers, the United States and the Soviet Union. Smith and Rathjens concluded that coercive policies might be justified in some cases, when all else failed, but they thought that success would "require both support from other supplier nations and sanctions going beyond constraints on nuclear trade." Non-plutonium fuel cycles were still worth promoting, they thought, but the emphasis in U.S. policy should be shifted from "unilateral denial" to multilateral efforts "relying on persuasion coupled with assurances of supply." [5]

Reagan's Approach to Non-Proliferation

When President Ronald Reagan took office, it was plain that he and his advisers would take a political approach to non-proliferation policy, with a strong emphasis on strengthening relations with allies and partners of the United States, and much less emphasis on arms control, sanctions, or export constraints. The Reagan team was strongly predisposed to think that the effort to curb the spread of technology had cost the United States sales in foreign markets and had bred resentment, without doing anything useful to restrain proliferation. During the 1980 campaign, Reagan at one point wondered out loud whether nuclear proliferation was "any of our business," when he was asked about Pakistan's efforts to acquire the technology needed to manufacture atomic bombs.

135

After taking office Reagan was careful not to appear complacent about proliferation. But he and his advisers held firm to the belief that friendly persuasion was better than arm-twisting, especially when dealing with countries they regarded as important to U.S. interests overseas. A couple of Reagan's appointees to positions relevant to non-proliferation policy had extensive business contacts in countries such as Japan, Taiwan, and South Korea.[6] No doubt the Reagan administration remembered that earlier Republican administrations had found a quiet approach effective in dealing with Asian client-states. During the Nixon-Ford years the U.S. government got South Korea and Taiwan to join the NPT and to back out of contracts to buy reprocessing plants from foreign suppliers.

When countries likely to go nuclear were considered important to U.S. security or U.S. interests, the Reagan approach was to gain influence by stressing U.S. reliability as a friend and supplier, and to use that influence — mainly in private, presumably — to discourage production and tests of atomic bombs. In the basic guidelines for non-proliferation policy that the administration released in July 1981 Reagan said: "We must establish this nation as a predictable and reliable partner for peaceful nuclear cooperation under adequate safeguards. . . . If we are not such a partner, other countries will tend to go their own ways and our influence will diminish."

The Reagan administration announced it would work out comprehensive long-term arrangements with Japan and the EURATOM countries, permitting spent fuel to be reprocessed without case-by-case U.S. approval. The administration renounced Carter's effort to discourage plutonium technologies such as breeder reactors, which depend on reprocessed fuels. The administration even talked of extracting plutonium from U.S. commercial reactors for use in weapons, but the Hart-Simpson Amendment of 1982 made such a move illegal.

In 1981-1982 the administration pushed through a sale of F-16 fighter planes to Pakistan even though it was known that Gen. Mohammed Zia ul-Haq, the country's dictator, was assembling the equipment and materials to build atomic bombs. The administration considered it essential to stay on good terms with Pakistan because of the Soviet Union's effort to pacify neighboring Afghanistan. As long as the United States sold Pakistan military equipment, Reagan officials reasoned, Washington would have the leverage to get General Zia to hold off on going nuclear. In 1983, for similar reasons, the Reagan administration facilitated nuclear exports to Argentina, India, and South Africa, even though all three countries were known to be intent on developing their capabilities to build atomic bombs.

Critics of the administration's approach contend that letting up on exports may help delay the day when other countries go nuclear, but that when that day comes, the critics believe, it will be easier for those countries to build their bombs quickly because of the equipment they have acquired. Privately, specialists who favor Reagan's approach often concede that this may indeed be the case. But as they see it, the spread of nuclear weapons cannot be stopped in the long run; it can only be slowed and managed.

This view is widely shared by some academic specialists on proliferation and even by former Carter officials. But it still raises hackles among activists such as Paul Leventhal, a former Capitol Hill aide who now heads a citizen group called the Nuclear Control Institute. The concern of his group, Leventhal said in an interview in September 1983, is to "prevent rather than manage the proliferation problem." [7]

State of the 'Regime'

From the historical record, it is obvious that the spread of nuclear weapons cannot be stopped easily. It also is clear, though, that barriers to proliferation are much stronger than many specialists expected them to be in 1974 when India tested its atomic bomb. At that time, what experts referred to as the "non-proliferation regime" seemed to be on the verge of collapse. The NPT had failed or seemed to have failed its first real test. Supplier rules and safeguards were full of holes. Because of the enormous increase in oil prices, many nations were preparing to acquire nuclear energy technology on a massive scale. It seemed that a great many countries soon would have the technology needed to build bombs, and there would be no firm moral or legal restraints capable of preventing them from doing so.

Some countries indeed have proceeded since May 1974 to acquire nuclear technology, and a few governments seem determined to go nuclear at the first opportunity. The number of countries has remained quite small, however, and those countries face much more formidable obstacles than they did a decade earlier. Non-proliferation agreements and understandings are much stronger, although still far from complete. Supplier rules and safeguards have been tightened. Economic pressures to acquire nuclear technology, especially sensitive fuel-cycle equipment,

India and Pakistan: . . .

India, of all the economically "less developed" countries, has by far the most advanced and elaborate nuclear program. It tested an atomic bomb in May 1974, and it was reported to be prepared to resume testing at any time. While it has not stockpiled atom bombs, as far as the United States knows, it is working on delivery systems suitable for carrying nuclear weapons. With extensive assistance from the Soviet Union, it has used missiles to launch satellites into orbit.

India has three atomic power complexes: two units at Tarapur built by U.S. firms; two Canadian CANDU units at Ranagratapsagar in Rajasthan; and two CANDU-like units it built itself near Madras. The power reactors are safeguarded, but India also has three unsafeguarded reprocessing plants that it built on its own and several unsafeguarded research reactors, including the 40 MW Cirus reactor at the Bhabha Atomic Research Center near Bombay. The Cirus reactor provided the plutonium for the 1974 bomb.

Shortly after India's 1974 test, Pakistan's ruler, Zulifika Ali Bhotto, said that his people would build atomic weapons as well, even if they had to "eat grass." Initially, the Pakistanis sought to acquire an atomic weapons capability by purchasing a reprocessing plant from France. After the Carter administration persuaded the French government to back out of the deal, the Pakistanis embarked on a radically different route: they set about getting all the components required to build a uranium enrichment plant of the centrifuge type, an extremely advanced enrichment technology that only came into commerical use in the 1970s, replacing the older gaseous diffusion method. Pakistan is thought to have acquired much of the information it needed to build the plant from a Pakistani scientist who worked at the URENCO enrichment plant at Almalo, Holland, from 1972 to 1975. URENCO is a British-German-Dutch

---

are much weaker than expected. And what Warren H. Donnelly, a senior specialist with the Congressional Research Service, has called "a general predisposition against nuclear weapons" is much stronger.[8]

Since May 1974 no country outside the nuclear club has openly tested an atomic bomb. India itself has not proceeded to build a nuclear arsenal, as many people expected it to. Nor has Pakistan, despite the

. . . Unlikely Users of Nuclear Technology

consortium that pioneered the commericalization of centrifuge enrichment.

Pakistan's unexpected decision to acquire an exceptionally complex technology caught nuclear officials in the supplier countries off guard, and it gave corporate executives an excuse to supply sensitive equipment on the ground that it had not been officially designated as sensitive. Swiss companies supplied especially critical components for Pakistan's enrichment plant. Claude Zangger, the Swiss official responsible for monitoring nuclear exports, who served as the first chairman of the London supplier group, said in an interview that the equipment obtained from Switzerland was not "sensitive," strictly speaking, but that it was "necessary" in the production of enriched uranium by the centrifuge method.

By the time the governments that are more committed to non-proliferation efforts got wind of Pakistan's enrichment program, work on the entrifuge plant was already quite advanced. Meanwhile, it was learned that the Pakistanis had proceeded on their own with the construction of a reprocessing plant, having obtained blueprints for such a facility from the French before the deal with France was cancelled. The plant would enable the Pakistanis to extract plutonium from fuel irradiated in a 140 MW power reactor that Canada finished building in 1972. While the reactor is safeguarded, the International Atomic Energy Agency (IAEA) has reported suspicious activities at the plant repeatedly in recent years. In 1982 the IAEA said it could not adequately verify safeguard compliance in two countries — and it generally was understood that the two countries were Pakistan and India.

Sources: K. Subrhamanyam, "The Indian Nuclear Explosion and its Impact on Security," *India Quarterly* (October-December 1974); WGBH Boston, "The Islamic Bomb," Nov. 5, 1980 and June 11, 1981 (update).

stated determination of both General Zia and his predecessor Zulifika Ali Bhotto to acquire nuclear weapons at the first opportunity. By early 1984 both countries were reported to have prepared test sites for atomic bombs, but both had come under close international scrutiny and so far, at least, an unrestrained nuclear arms race on the subcontinent had been averted.

South Africa

South Africa, which has large uranium reserves, announced plans to build a commercial uranium enrichment plant in 1975. The South Africans bought an unusual kind of enrichment technology, which was developed by Dr. E. W. Becker at a West German nuclear research center in Karlsruhe, West Germany, from the German firm STEAG in 1973. Becker's technology has not been considered suitable for commercial use by companies or governments in Germany or any other countries except South Africa. This has fed speculation that South Africa's real interest in the technology is military.

In August 1977 the Russian and French governments charged that South Africa was preparing a site for a nuclear test. President Jimmy Carter extracted assurances from South Africa that no test would take place, but in the following years the South African government steadfastly refused to join the Non-Proliferation Treaty or open all its facilities to international inspectors. Since the fall of 1979, when a U.S. surveillance satellite spotted a mysterious flash in the South Atlantic, there has been lively speculation about whether South Africa tested an atomic or hydrogen bomb, perhaps in collaboration with Israel. Military planners in South Africa and Israel would have reason to be interested in acquiring tactical nuclear weapons, and the two countries could contribute complementary resources to a development program. The Israeli and South African governments are thought to have been cooperating in atomic research at least since 1976, when Ampie Roux — the top person in South Africa's nuclear program — made several trips to Israel.

For further information see Barbara Rogers and Zdenek Cervenka, *The Nuclear Axis: Secret Collaboration Between West Germany & South Africa* (New York: Times Books, 1978).

To be sure, one secret nuclear weapons test may have occurred since May 1974. On Sept. 22, 1979, a U.S. surveillance satellite over the South Atlantic detected a flash with the characteristic "signature" of a nuclear weapons test. The U.S. government has never publicly confirmed that it believes a nuclear test took place, and some independent weapons experts believe there was no test. Officials have not been eager to discuss the incident in the South Atlantic because U.S. relations are sensitive and

Israel

As early as 1968, the CIA had informed the president of its opinion that Israel had assembled an arsenal of nuclear weapons. In 1978 the CIA released a 1974 memorandum stating the agency's belief "that Israel already has produced nuclear weapons." It generally is assumed that bomb work has been done at Israel's unsafeguarded Dimona complex, which includes a research reactor that France supplied in the 1950s, and a reprocessing plant. Israel also may have bought or stolen nuclear materials from private companies in the United States and Europe.

On Feb. 21, 1980, CBS News reported that Israel had tested a nuclear weapon in the South Atlantic the preceding fall, probably in collaboration with South Africa. The Israeli government immediately denied the charge and expelled Dan Raviv, the correspondent who filed the story. The CBS News management, expressing confidence in Raviv, made him head of their London bureau.

Since 1980 intelligence specialists have reached no firm conclusions about whether there was a nuclear test in the South Atlantic or, if there was, what kind of test it was and who was responsible for it. Satellite photos and seismographic information strongly suggested that a test took place, but no evidence of fallout has been detected. Nuclear weapons experts have speculated that there may have been a test of a neutron bomb, a special kind of hydrogen bomb designed for battlefield use. Israel is thought to have considerable expertise in the area of bomb design, and South Africa has the technology to produce enriched uranium, which is superior to plutonium in hydrogen but not atomic bombs.

Sources: Howard Kohn and Barbara Newman, "How Israel Got the Bomb," *Rolling Stone* (December 1977); John J. Fialka, "How Israel Got the Bomb," *Washington Monthly* (January 1979).

controversial with both of the countries most widely suspected of having tested a weapon — South Africa and Israel.

Whether or not South Africa and Israel actually have built atomic bombs, leaders of the two countries clearly believe it would be a grave mistake to offend world opinion with aggressive statements about their nuclear plans. While they refuse to say that they never will test or use atomic bombs, and while they occasionally hint they already may have

Argentina and Brazil . . .

South America's two largest countries often are described as being in a "race" to build atomic bombs, but so far Argentina has been doing most of the running because of its earlier start. For many years Argentine governments have been assembling a complete nuclear fuel cycle based on reactors fueled by natural rather than enriched uranium. By all accounts, the natural-uranium fuel cycle is the most efficient and economical route to an atomic weapons capability.

Argentina currently has two nuclear power plants fueled by natural uranium, a 320 MW plant built by the German company Kraftwerk Union (KWU) and a 664 MW CANDU reactor supplied by Atomic Energy of Canada Ltd. These reactors are under international safeguards, but Argentina also has built a number of small research reactors, which are not safeguarded. It also has experimented with reprocessing technology on a small scale; a larger reprocessing facility, which Argentina has not agreed to open to IAEA inspection, has been scheduled to come into operation around 1987. In November 1983, shortly before the inauguration of President Raul Alfonsin after seven years of military rule, the head of Argentina's Atomic Energy Commission announced that the AEC had successfully developed an indigenous uranium enrichment technology. The Argentine AEC also has built uranium milling plants with French assistance and a yellowcake factory where uranium ore is converted to uranium dioxide with German help.

In 1980, against the wishes of the Carter administration, Switzerland sold Argentina a processing plant to manufacture heavy water. When the plant is completed around 1985, Argentina will be able to operate its power reactors without relying on foreign suppliers for heavy water, which would make it much easier to renounce international safeguards the country

nuclear arsenals, they avoid discussing the subject as much as possible. South Africa's uranium enrichment complex, which could produce the fissionable materials for atomic or hydrogen bombs, is called "Valindaba" — in the African language Sotho, the place where "we don't do much talking."

Leaders of near-nuclear countries do not advertise their interest in atomic bombs. Generally, they prefer to cultivate a certain ambiguity

... Competing to Build Bombs

currently accepts. When the heavy water plant was bought, the Argentines also purchased a 740 MW natural uranium reactor from KWU, and many observers suspected collusion between the German and Swiss suppliers. Canada had offered to sell Argentina a CANDU reactor and heavy water plant, but only if the Argentines accepted full-scope safeguards. Argentina got similar facilities from Germany and Switzerland, at a much higher price, but it did not have to accept comprehensive safeguards.

The same kind of thing happened in 1974-1975 when Brazil decided to purchase a complete nuclear fuel cycle from foreign suppliers. Bechtel, a U.S. engineering firm, originally made an offer but was forced to withdraw it by the Ford administration. Germany stepped into the breach and made a deal to supply Brazil with up to eight nuclear power plants, a reprocessing facility, and enrichment technology.

Some Brazilian scientists like Jose Goldenber, who currently is the head of Sao Paulo's main electrical utility, were critical of the deal with Germany because they thought that reliance on nuclear power would be unnecessary and wasteful in light of Brazil's vast hydroelectric resources. In the United States and Europe, many people thought it was irresponsible for West Germany to export sensitive technologies. Although the Carter administration's efforts to cancel the deal were unsuccessful, work has lagged far behind schedule on the first two reactors, and Brazil will not be able to buy additional reactors any time soon because of financial problems. Lacking the reactor orders they expected, the Germans may refuse to supply the enrichment and reprocessing technology.

Sources: Norman Gall, "Atoms for Brazil, Dangers for All," *Foreign Policy* (Summer 1976); Hartmut Krugmann, "The German-Brazilian Nuclear Deal," *Bulletin of the Atomic Scientists* (February 1981).

about their intentions. Thus they are able to play for time, minimize international criticism, and — with skill — extract favors from other countries in exchange for continued abstinence. This is the kind of game that has helped keep a restrained but vigorous nuclear rivalry between Argentina and Brazil alive. Both countries have been systematically acquiring complete nuclear fuel cycles, including the sensitive enrichment and reprocessing technologies needed to make bombs. Both countries have signed the Treaty of Tlatelolco, a 1967 agreement that bars the introduction of nuclear weapons into Latin America. But Argentina has refused to ratify the treaty, and Brazil — along with Chile — has specified that its ratification will take effect only when all other countries in the region have joined. Cuba is the one country in Latin America that has refused to sign or ratify the Tlatelolco agreement. During the Carter years, Cuba is said to have told Washington that it would be willing to join the accord if relations with the United States were normalized, the U.S. trade embargo was lifted, and agreements were reached about military maneuvers in the Caribbean.

Between 1976 and 1983, when Argentina was ruled by a military junta, the government repeatedly refused to give Washington assurances about its nuclear intentions. At the end of 1983, however, when a civilian government returned to power in Buenos Aires, the chances of getting Argentina to accept comprehensive safeguards seemed to be significantly improved. Hans Blix, director general of the International Atomic Energy Agency (IAEA), went to Buenos Aires immediately after the inauguration of Raul Alfonsin, Argentina's new president. In a letter to *The Washington Post* on Dec. 13, 1983, Blix reported that "discussions with Argentina about comprehensively safeguarding the Argentinian nuclear program under the Tlatelolco Treaty" had been "very active in the last nine months." U.S. Vice President George Bush attended Alfonsin's inauguration, and U.S. officials said at the time that "their priority" was to "nudge the Argentines gently to put their growing nuclear program under international safeguards." [9]

The stubborn refusal of countries most likely to go nuclear to sign and ratify non-proliferation agreements has been the most frustrating aspect of the international effort to stop the spread of atomic bombs. But the number of near-nuclear countries has not grown significantly since May 1974, and such countries increasingly are isolated and exposed. When India tested its bomb it was just one of a great many countries that refused to join the NPT. Then, the Non-Proliferation Treaty had just 83 members; 10 years later, at the beginning of 1984, it had 120.

A few dozen countries remain outside the NPT, but the most important of them are beginning to stand out in sharp relief when parties and non-parties are charted on a global map. While the near-nuclear countries are still trying hard to acquire sensitive technology such as reprocessing and enrichment equipment, their efforts are more likely to be detected and deterred than ever before. Even when they do manage to

Nuclear Free Zones

The Treaty for the Prohibition of Nuclear Weapons in Latin America, generally known as the Treaty of Tlatelolco, laid the foundation for the world's first nuclear free zone. The treaty was adopted at Mexico City in 1967 following several years of negotiations in which Mexico's Alfonso Garcia Robles played a leading role. Because of his contributions to United Nations disarmament negotiations, Garcia Robles was awarded the Nobel Peace Prize in 1982, along with Alva Myrdal, Sweden's leading disarmament specialist.

Parties to the Tlatelolco Treaty, like parties to the Non-Proliferation Treaty, agree to use nuclear energy exclusively for peaceful purposes. But the Latin American treaty provides for a stricter control system. Like the NPT, the Tlatelolco agreement calls for verification of non-proliferation pledges by the International Atomic Energy Agency. But the agreement also relies on a special organ, the Agency for the Prohibition of Nuclear Weapons in Latin America (OPANAL), which has the authority to conduct inspections in member countries whenever a charge is brought that a party may be violating its commitments.

As of the beginning of 1984, 22 Latin American countries have become parties to the Tlatelolco treaty, leaving just four nations — Argentina, Brazil, Chile, and Cuba — outside the agreement. In addition, all five countries with arsenals of nuclear weapons have ratified the treaty's Protocol II, in which they promise "Not to use or threaten to use nuclear weapons" against full parties of the treaty. England, the Netherlands, and the United States have ratified Protocol I, which bars them from introducing nuclear weapons onto the soil of their Latin American possessions. France, the only other outside country with territorial interest in Latin America, has signed but not ratified Protocol I.

Sources: Alfonso Garcia Robles, "The Latin American Nuclear-Weapon-Free Zone," The Stanley Foundation, Occasional Paper 19, 1979; John R. Redick, "The Tlatelolco Regime and Nonproliferation in Latin America," in *Nuclear Proliferation*, ed. George H. Quester (Madison: University of Wisconsin Press, 1981) and Jozef Goldbat, "Tlatelolco and the Falklands," *Bulletin of the Atomic Scientists* (May 1983).

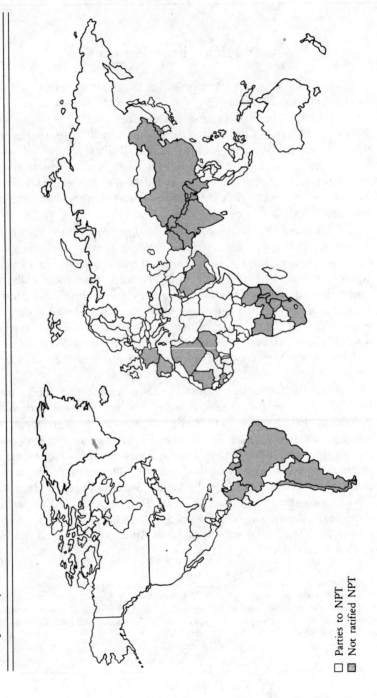

Membership in Non-Proliferation Treaty
as of January 1, 1984

□ Parties to NPT
▨ Not ratified NPT

Source: U.S. Arms Control and Disarmament Agency

get added items, they are almost sure to come under heavy international pressure not to proceed with bomb tests. This was true of Pakistan, for example, after it acquired sensitive nuclear equipment in the early 1980s.

The list of countries with unsafeguarded facilities has not grown since 1974, and it is more and more difficult for additional countries to get sensitive equipment. In the immediate wake of the Arab oil embargo of 1973-1974, an especially alarming trend arose that showed the willingness of some European governments to permit the sale of sensitive technology to "sweeten" big reactor deals with non-nuclear countries. West Germany negotiated a multibillion-dollar contract to supply Brazil with a complete nuclear fuel cycle, including reprocessing and enrichment plants. France agreed to sell Pakistan a reprocessing plant, and in Iraq the French began to build a large research reactor designed to run on highly enriched uranium — a material usable in atomic bombs.

In the mid-1970s a number of writers proposed the establishment of an international market-sharing arrangement in which the major reactor manufacturing countries would be provided with assured proportions of world reactor sales in exchange for commitments not to sell sensitive equipment.[10] As it happened, the intense commercial rivalries among the major manufacturing nations made such an arrangement workable. But in the meantime it also became apparent that the creation of a nuclear cartel was unnecessary. France backed off quietly from its agreement to sell reprocessing technology to Pakistan, and in December 1976, just before Carter took office, the French announced that they were suspending all further sales of reprocessing technology. In June 1977 the Federal Republic of Germany announced that it, too, would make no further sales of reprocessing equipment.

Since the mid-1970s there have been no new sales of reprocessing or enrichment equipment to non-nuclear countries. Reactor sales also have lagged dramatically behind the mid-1970s projections. In Iran, which concluded big nuclear deals with West Germany and France in 1976 and 1977, all reactor construction came to a halt in 1970 after the shah was overthrown in an "Islamic revolution." Germany's Brazil deal has proved to be a one-of-a-kind transaction, and even it may be coming unglued. In all but a very few Third World countries, sagging economic growth and acute debt problems preclude large purchases of expensive nuclear technology in the near future. Only South Korea and Taiwan have sizable reactor programs; in Mexico and the Philippines, work is proceeding at a slow pace on the countries' initial reactor projects.

147

Unsafeguarded Facilities
December 1982

Country	Facility	Indigenous or imported	First year of operation
Argentina	(Reprocessing plant at Ezeiza)	Indigenous	1956
India	Apsara research reactor	Indigenous	1960
	Cirus research reactor	Imported (Canada)	1972
	Purnima research reactor	Indigenous	1960
	Fuel fabrication plant at Trombay	Indigenous	1974
	Fuel fabrication plant, CANDU-type fuel elements, at the Nuclear Fuel Cycle Complex, Hyderabad	Indigenous	
	Reprocessing plant at Trombay	Indigenous	1964
	Reprocessing plant at Tarapur	Indigenous	1977
Israel	Dimona research reactor	Imported (France/Norway)	1963
Israel	Reprocessing plant at Dimona	Indigenous (in cooperation with France)	
Pakistan	Fuel fabrication plant at Chashma	Indigenous (in cooperation with Belgium)	1980
	(Reprocessing plant near Islamabad)	Indigenous (drawing on design information furnished by Belgo-Nucleaire)	Status unknown
	(Uranium enrichment facility at Kahuta)	Indigenous (based on import of technology and equipment from a variety of countries)	Status unknown
South Africa	Enrichment plant at Valindaba	Indigenous (in cooperation with FR Germany)	1975

Source: *SIPRI Yearbook,* 1983.

In the years immediately after India's 1974 test, it was widely believed that a great many countries would soon have so much nuclear materials and equipment that they would be able to build bombs in a matter of days or weeks. Without violating the letter of agreements or circumventing safeguards, the argument went, countries could all but go nuclear, and, almost inevitably, many of them would do so.

The notion that technology would spread inexorably and drive increasing numbers of countries into the nuclear club has turned out to be almost completely wrong. Technology has not spread so very inexorably, and, even when it has spread, it has not prompted countries to proceed with atomic bomb tests. In the mid-1970s, U.S. legislators paid close attention to graphs that seemed to predict an exponential increase in the numbers of countries that would acquire nuclear technology and join the nuclear club. Such graphs have proved to be misleading.

Disputed Adequacy of Safeguards

Among U.S. specialists on non-proliferation there currently is wide agreement that prospects for preventing the spread of nuclear weapons are reasonably good. In a recent collection of essays by leading experts, George H. Quester noted in his editorial introduction that all the contributors rejected "the conclusion that halting the spread of nuclear weapons would be impossible." At worst, Quester said, "further proliferation may come much more slowly than the pessimists have predicted and may come in more ambiguous or diluted forms." [11] To the extent proliferation takes place, Quester seemed to be saying, countries are likely to deny that they have weapons and to be unlikely to test or use them except under the most extreme circumstances.

That may be so, but experts also are inclined to agree that near-nuclear countries will keep their ambitions in check only if nefarious activities are likely to be detected and punished. Opinion among experts is quite divided about whether the international safeguards system is adequate to these tasks. Even the people who are responsible for managing and implementing safeguards wonder whether violations of the NPT or bilateral agreements always can be detected in time for action to be taken. They are still more skeptical about whether action would be taken if violations were detected.

As of February 1984 about 150 IAEA inspectors were responsible for monitoring the whereabouts of some 15,000 tons of nuclear material

Nuclear Proliferation as Projected in the Mid-1970s

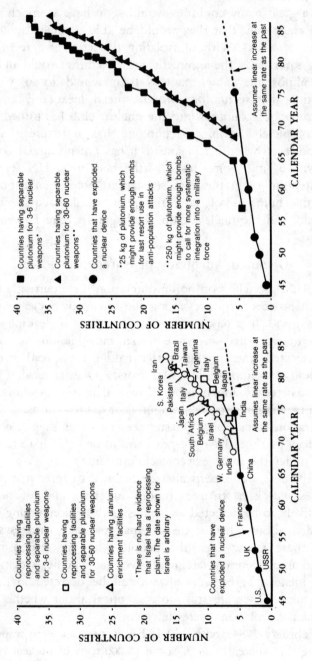

With charts that showed how the spread of nuclear technology would enable more and more countries to build atomic weapons, scholars such as Albert Wohlstetter sought to highlight the dangers connected with the export of enrichment and reprocessing technologies. But such charts tended to leave the impression that nuclear weapons would spread much more than they actually have. The illustration to the left here is from Albert Wohlstetter et al., "Moving Toward Life in a Nuclear Armed Crowd," a report prepared for the U.S. Arms Control and Disarmament Agency, April 22, 1976; the second is from Albert Wohlstetter, "Spreading the Bomb Without Quite Breaking the Rules," *Foreign Policy* (Winter 1976-1977).

worldwide. That is easily enough material to make thousands of atomic bombs, and it was stored at approximately 350 sites in more than 50 countries. The IAEA operates on an annual budget of less than $100 million.

In March 1981, Emanuel Morgan — a former safeguards inspector — informed the U.S. Nuclear Regulatory Commission (NRC) that the IAEA was "incapable of detecting the diversion of a significant quantity" of nuclear fuel "in any state with a moderate to large nuclear energy establishment." [12] The same year, specialists at Batelle's Pacific Northwest Laboratory informed the NRC that the IAEA would need to increase its inspection effort tenfold to do its job adequately. "I don't think most people realized how serious the problem was," Leonard Weiss, minority staff director of the Senate Subcommittee on Energy, Nuclear Proliferation, and Government Processes, commented at the time. Weiss, an aide to Sen. John Glenn, D-Ohio, was a principal author of the 1978 Non-Proliferation Act.[13]

In the United States, which is responsible for keeping track of its own nuclear materials without much help from the IAEA, significant quantities of plutonium and highly enriched uranium have been reported missing in recent years. An Energy Department audit in 1982 found that 55.6 pounds of plutonium could not be accounted for in inventories taken from Oct. 1, 1980, to May 31, 1981 — enough material for a half dozen atomic bombs. In 1977 the federal government found that a cumulative total of more than 8,000 pounds of highly enriched uranium and plutonium had disappeared since the beginning of the nuclear era. Many rumors and reports indicated that some of this missing material might have found its way to Israel, where it may have been used to produce atomic bombs.

In late 1980 a suspicious pattern of events began to unfold in Baghdad, Iraq, where the French were completing a large research reactor for the Iraqi government. In September, Iraq temporarily evicted all French technicians, saying it could not protect them because of its war with Iran, which had broken out earlier that year. Two months later, Iraq's government announced it would not permit IAEA inspection of its nuclear facilities until the war with Iran ended.[14] Meanwhile, the French government was coming under mounting international pressure to modify or cancel its agreement with Iraq. The Iraqi government refused French entreaties to allow it to substitute less highly enriched uranium rather than bomb-grade uranium for the research reactor. Iraq did agree, however, to allow the French a presence at the plant for a number of years after it was to come into operation.

On June 7, 1981, squadrons of Israeli F-15 and F-16 fighters totally destroyed Iraq's reactor in an attack of remarkable surgical precision. It was the first direct military action any country had ever taken to prevent another country from going nuclear, and it unleashed a storm of controversy about Israel's justification for the attack, the integrity of the IAEA system, and the non-proliferation regime as a whole.

At the IAEA the raid was treated as a direct assault on the agency. Less than a week after the raid, the IAEA's board of directors voted that Israel should be considered for suspension from the agency. Later in the year the IAEA's members voted narrowly to deny Israel credentials to attend the next agency meeting. This move provoked the United States to temporarily suspend funding for the IAEA, and it was taken by agency officials as a threat to the survival of the safeguards system. Since then, U.S. funding has been restored and so have Israel's credentials.

Critics of the IAEA were quick to argue that the Israeli raid demonstrated the inadequacy of the international safeguards system. An American inspector, Roger Richter, resigned from the IAEA before the raid because he felt that his suspicions about Iraq were not being treated with sufficient respect. And after Israel's raid he told Congress that Iraq's pattern of activities left him with little doubt that the country's rulers were planning to build nuclear weapons.[15] Richter said that Iraq could have used the reactor to convert unsafeguarded, unenriched uranium into plutonium and could have refined the plutonium for a bomb in an unsafeguarded "hot cell" laboratory that Italy was building at the reactor complex. Richter pointed out that Iraq was procuring a large stockpile of unenriched uranium for which there was no apparent use.

Not everybody agreed with Richter's assessment of Iraq's intentions. Senate aide Weiss, for example, believed that Iraq did not plan to use the French research reactor to produce plutonium because it was not ideally suited for that purpose; Weiss thought that Iraq was hoping to purchase a large natural uranium reactor to produce plutonium.[16] However that may be, even if Richter was completely correct in his assessment the implications of his testimony were open to conflicting interpretations. In a sense it was Iraq's apparent effort to exploit several loopholes in the safeguard system that made Israel and the international community suspicious. In this sense the raid may have vindicated the IAEA. Without the

How Iraq Might Have Used Safeguarded Reactor to Produce Bombs

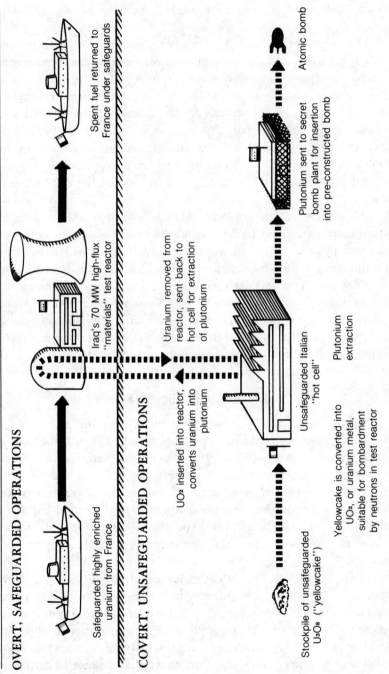

OVERT, SAFEGUARDED OPERATIONS

Safeguarded highly enriched uranium from France

Iraq's 70 MW high-flux "materials" test reactor

Spent fuel returned to France under safeguards

COVERT, UNSAFEGUARDED OPERATIONS

UO₂ inserted into reactor, converts uranium into plutonium

Uranium removed from reactor, sent back to hot cell for extraction of plutonium

Stockpile of unsafeguarded U₃O₈ ("yellowcake")

Yellowcake is converted into UO₂, or uranium metal, suitable for bombardment by neutrons in test reactor

Unsafeguarded Italian "hot cell"

Plutonium extraction

Plutonium sent to secret bomb plant for insertion into pre-constructed bomb

Atomic bomb

Based on testimony to Congress by Roger Richter, a former International Atomic Energy Agency inspector.

153

agency's safeguard system, there might have been no warning signals of any kind.

To people who have been studying nuclear proliferation for many years, Israel's Iraq raid highlighted what may be the single most important weakness in the world's non-proliferation regime — the absence of any agreed-upon actions to be taken by the international community if safeguard violations are detected. When Iraq ejected technicians and inspectors, no international response occurred, even though the action was a clear and unequivocal signal that something was amiss. In the end, ironically, it was Israel — a country that may have nuclear weapons and has not joined the NPT — that acted to enforce the NPT's basic principle.

Of course Israel completely lacked legal or moral standing to take action against Iraq, and IAEA officials had no choice but to object in the strongest possible terms to the raid. Even if Israel had been certifiably non-nuclear and an NPT member in good standing, its legal right to take unilateral action against Iraq would have been highly uncertain. But so long as there is no agreed-upon international mechanism for enforcement of safeguard agreements, there may be no alternative to unilateral acts in the type of situation that Israel faced. In the current system, "even if the alarms sound, the fire engines have flat tires," Warren Donnelly said in an interview shortly after the Israeli raid.[17]

Proliferation in the Middle East

Since safeguards first were established in the late 1950s, there has been grave concern that countries would use membership in the IAEA system as a cover for efforts to acquire atomic bomb technology. Increasingly, that concern centers on states in the Middle East, an explosive region where none of the nuclear weapons states has more than a tenuous influence and where anything might happen. Apart from Pakistan and Israel, most of the countries in this region are parties to the NPT, but the sincerity of several is open to question.

Col. Muammar Qaddafi, the leader of Libya, an NPT party, has made little secret of his interest in nuclear weapons. According to reputable sources, he tried to purchase an atomic bomb from the People's Republic of China shortly after seizing power in 1969. In 1976 Qaddafi said: "A few years ago, we could hardly manage to procure a squadron of fighter planes. Tomorrow, we shall be able to buy an atomic bomb and all its component parts. The monopoly of the atom will be broken any day now." [18]

For some years there have been rumors and poorly substantiated reports that Qaddafi is helping to finance the construction of an "Islamic bomb" by Pakistan. Zulifika Ali Bhutto, General Zia's predecessor as leader of Pakistan, once wrote: "There's a Hindu bomb, a Jewish bomb and a Christian bomb. There must be an Islamic bomb." [19] Saddam Hussein, the president of Iraq, an NPT party, echoed such sentiments in an address to his Cabinet shortly after the Israeli raid in June 1981: "I think that any country . . . that has a positive responsibility toward humanity and peace must say to the Arabs: 'Here, take arms and face the Zionist atomic threat so you may prevent the Zionist entry [Israel] from using the atomic bomb against the Arabs and spare the world the danger of using atomic bombs in wars.' "

The day after Saddam Hussein made these statements, former Israeli defense minister Moshe Dayan was quoted in *The New York Times* as saying that while the Israelis "don't have any atomic bomb now," they have the capacity to construct one "in a short time." [20] Israel's official position is that it will not be the first country to introduce nuclear weapons into the Middle East. But the meaning of this policy is ambiguous. It could be taken to imply that Israel literally will not "go nuclear" unless some other country in the Middle East does so first. It also could be interpreted as a mere pledge not to be the first nation to use nuclear weapons in a Middle Eastern war.

Mohammed Heikal, an Egyptian journalist who is highly regarded in the West, expressed grave misgivings about the nuclear developments in the Middle East in a television interview in April 1981. "There will be a nuclear moment of truth in the Middle East," he said. "When I look to what can come . . . I really shiver." [21] The best that can be said of the Middle Eastern situation, perhaps, is that it has made the nations of the world even more attentive to the dangers of nuclear proliferation and more united in their determination to take concerted counter-measures.

Developments in Nuclear Club

In July 1983 Soviet foreign minister Andrei Gromyko said at the second special session of the U.N. General Assembly on Disarmament that the Soviet Union would be willing to place some of its peaceful nuclear facilities under international safeguards. This made the Soviet

Union the fourth nuclear weapon state — after the United States, England, and France — to place some civilian facilities under safeguards.

France and China still refuse to join the Non-Proliferation Treaty, but they continually are edging closer toward membership. After the socialist government of Francois Mitterand took office in 1981, some of his top foreign policy advisers indicated that French adherence to the NPT and the partial test ban treaty was under consideration. However, a government statement on arms control, which supposedly was being prepared in early 1982, did not materialize.[22] Instead, Mitterand became increasingly preoccupied by the spread of a grass-roots peace movement in Germany, the Netherlands, and England, which he regarded as a threat to the cohesion of the West's defenses.

From the perspective of international non-proliferation efforts, the vehemence and growth of the peace movement is an important development. In Germany and England, groups with substantial popular support argue for removing all nuclear weapons from the two countries. Germany's "Greens," a party that won more than 5 percent of the vote, the amount required to gain seats in parliament in the March 1983 election, takes the position that no nuclear weapons should be introduced into the Federal Republic and that existing tactical nuclear warheads should be removed. England's Labor party now favors dismantling the country's independent nuclear deterrent. The positions taken by the Greens and the Labor Party are not supported by majorities in either country, but they are significant in that they show that people are beginning to talk about reversing the spread of nuclear weapons — not just limiting it.

In the United States there is little support for unilateral disarmament measures, but in 1981-1982 a movement to freeze the testing, production, and deployment of nuclear weapons began to catch fire at the grass-roots level. The proposal favored by the freeze movement contains essentially the same suggestion that the Indian government made in non-proliferation negotiations during the 1960s and 1970s. As yet, however, the freeze movement has had little direct influence on U.S. arms control policy.

To a great extent the European peace movement and the U.S. freeze movements have fed off developments that augur poorly, on the whole, for the success of non-proliferation efforts: 1) the sharp deterioration in U.S.-Soviet relations, which cannot help but undermine the joint policies that produced the safeguard system and the NPT; 2) the increased deployment of tactical nuclear weapons in both West and East Europe; and 3) the growing currency of "nuclear-war-fighting doctrines" based on the notion that it might, after all, be possible to "win" a nuclear war.

If preparations for fighting tactical nuclear wars in Europe go far enough, it is feared that countries such as Sweden, Switzerland, and Yugoslavia may begin to have second thoughts about their membership in the NPT. All three countries put a great deal of money and effort into their conventional defenses, and their operating assumption always has been that they would be able to put up a good fight against a conventional attack. If the impression grows that the nuclear weapons states would use tactical nuclear warheads in a European war as a matter of course, people in the smaller European countries could begin to think that they have no recourse but to acquire such weapons, too.

Pros and Cons of Absolutist Positions

In the freeze and peace movements, a great many people are inclined to believe that the only answer to the proliferation of nuclear weapons is to bar all uses of nuclear energy. In a world without nuclear power plants and fuel cycle facilities, it would indeed be harder to get the material and equipment for atomic bombs. But the possibility of nuclear proliferation would not be completely eliminated. The knowledge needed to build atomic bombs would still be with us, and a country or terrorist group always would have some chance of building a small plutonium production reactor and a reprocessing plant to get the materials for bombs. Studies done for the Congressional Research Service in the mid-1970s indicated that a reactor capable of producing the plutonium for one bomb a year could be built for $13-$26 million, and that a reprocessing plant could be built for less than $25 million.[23]

Because of the danger that some malicious country or group could still build a bomb secretly, even if the world totally disarmed and eliminated all nuclear weapons facilities, some scholars have concluded that it is a mistake to think that it ever will be possible to get rid of nuclear weapons completely. In their view, if we disarmed completely we would be putting ourselves in a position to be terrorized by any government or group that managed to get its hands on one or two bombs.

For now it is perhaps best to remember that significant strides have been made in combatting the proliferation of nuclear weapons, even though progress has been slow and halting. Bertrand Goldschmidt, a leading figure in France's nuclear program, considers the progress made so far nothing short of miraculous. The first miracle, he told *The Christian Science Monitor* in 1981, is that "safeguards were accepted — a political revolution!" [24]

"Miracle No. 2," he said, has been the declining rate of proliferation in the 40 years since the nuclear age began. "In the first decade after World War II, three countries got the bomb: the U.S. in 1945, the U.S.S.R. in 1949, the United Kingdom in 1952. In the second decade, two: France in 1960, China, 1964. In the third decade, one: India, 1974. In the fourth decade, none — if the flash in the sky on Sept. 22, 1979 between South Africa and Antarctica was not a bomb. . . . The fact is that no one has embarked on an open nuclear arms campaign since 1964."

The third miracle, Goldschmidt said, has been that "five countries that could easily have built a bomb after the war did not" — West Germany, Japan, Italy, Canada, and Sweden. Those countries, he might have added, made themselves the core of a non-nuclear club, and it is far from impossible that this club will keep growing in the decades to come.

Notes

1. See Richard K. Betts, "Paranoids, Pygmies, Pariahs and Non-Proliferation," *Foreign Policy* (Spring 1977), 157-181, for an argument against taking a moralistic and universalistic approach to proliferation policy. Also see Lewis A. Dunn, *Controlling the Bomb: Nuclear Proliferation in the 1980s* (New Haven: Yale University Press, 1982).
2. Stanley Hoffman, "Nuclear Proliferation and World Politics," in *A World of Nuclear Powers?*, ed. Alastair Buchan (Englewood Cliffs, N.J: Prentice-Hall, 1966), 89-90.
3. Seymour M. Hirsch, *The Price of Power* (New York: Summit Books, 1983). The source of this quote is Morton Halperin who resigned in protest against Kissinger's policies and subsequently sued Kissinger for putting wiretaps on his phone.
4. Ted Greenwood and Robert Haffa Jr., "Supply-Side Non-Proliferation," *Foreign Policy* (Spring 1981): 125.
5. Gerard Smith and George Rathjens, "Reassessing Nuclear Non-Proliferation Policy," *Foreign Affairs* (Spring 1981): 875-894.

6. This was true of Richard Allen, Ronald Reagan's first national security adviser, and James L. Malone, who served as assistant secretary of state for the environment and scientific affairs.

7. The Nuclear Control Institute, a group supported by members and foundations, lobbies for stricter anti-proliferation legislation and brings legal actions with other "public interest groups" to ensure strict enforcement of existing rules.

8. See Warren H. Donnelly, "Changing Pressures on the Non-Proliferation Regime," in *World Armaments and Disarmament: 1983 Yearbook,* Stockholm International Peace Research Institute, 69. Donnelly, a senior specialist with the Congressional Research Service, is one of the most knowledgeable people in the United States on the subject of nuclear proliferation.

9. See *The New York Times,* Dec. 12, 1983.

10. See Abraham A. Ribicoff, "A Market-Sharing Approach to the World Nuclear Sales Problem," *Foreign Affairs* (July 1976), 763-803. During the first three months of 1977 the author interviewed nuclear industry executives and government officials in Germany, Sweden, England, and France about the merits of the Ribicoff proposal, which was drafted by Paul Leventhal. The author found little or no interest in the idea in Germany and France, where executives and officials were confident that they would do better in the world market without cooperating with U.S. counterparts. There was some slight interest in the idea of a cartel only in England, where the industry was in deep trouble.

11. See George H. Quester, "Introduction," in *Nuclear Proliferation: Breaking the Chain* (Madison: University of Wisconsin Press, 1981), 1.

12. Quoted in *The New York Times,* Dec. 22, 1981.

13. Ibid.

14. See William Sweet, "Iraq's Bombshell Announcement," *Editorial Research Reports, Vol. II, 1980* (Washington, D.C.: Congressional Quarterly, 1981), Nov. 20, 1980, and "Nuclear Alarms Ringing," *Editorial Research Reports, Vol. I, 1981* (Washington, D.C.: Congressional Quarterly, 1981), May 8, 1981.

15. See Roger Richter, "Testimony," *The Bulletin of the Atomic Scientists,* (October 1981): 29-31.

16. Quoted in William Sweet, "Controlling Nuclear Proliferation," in *Editorial Research Reports, Vol. II, 1981* (Washington, D.C.: Congressional Quarterly, 1982), 515.

17. Ibid., 516.

18. Ibid., 517.

19. Ibid., 519.

20. Quoted in *The New York Times,* June 24, 1981.

21. Quoted on an ABC-TV feature about the spread of nuclear weapons in the Middle East on April 27, 1981.

22. Jacques Huntzinger meeting with reporters at the American Enterprise Institute on April 15, 1982, led his audience to expect such a paper. Huntzinger is a high-level adviser to Francois Mitterand.
23. See the papers by John R. LaMarsh in U.S. Congress, *Nuclear Proliferation Factbook* (Washington, D.C.: Government Printing Office, 1977), 501-585.
24. See *The Christian Science Monitor,* Dec. 4, 1981.

Part III

The Arms Race

KER

2 FUEL TRAILERS

METALLURG ANOSOV

RECTOR

DIVNOGORSK

BRATSK

On November 2, 1962, American reconnaissance planes took this photograph of the Mariel Naval Port in Cuba. The photograph confirmed that nuclear missiles were being loaded onto ships after the Soviet Union promised to remove the weapons from the island.

Chapter 9

NUCLEAR CONFRONTATIONS AND CONTROL

The bombing of Hiroshima and Nagasaki established that nuclear weapons could be built and that they could be used to accomplish political ends. At the time the bombs were dropped, it was plain that the United States and the Soviet Union would be the dominant powers in the postwar world, and relations between the two countries already were deteriorating rapidly. Perhaps inevitably, the suspicion arose that one country might use nuclear weapons against the other.

Three times in less than 150 years the Soviet Union had been invaded and devastated by armies from the West; World War II alone had cost the Soviet Union more than 18 million lives and left the most highly industrialized part of the country in ruins. Soviet leaders, whose ideology portrayed communism and capitalism as mortal enemies, unanimously supported Stalin's policy of establishing an Eastern European buffer zone against renewed invasion from the West. While Russia had been allied with Western democracies in the struggle against Hitler, Soviet leaders had not forgotten that the United States intervened in their civil war after the Revolution of 1917, and they were well aware of American anti-communist sentiment.

In the United States not everybody regarded a military collision between capitalism and communism as inevitable, but there were good reasons to be worried about Stalin's brand of communism. Because of the reign of terror he had instituted in the 1930s, Stalin was regarded as a tyrant similar to Hitler, and many Americans were inclined to think in the wake of World War II that tyrants, by definition, were international aggressors. Even though U.S. losses in the war had been small by comparison with the Soviet Union's, Pearl Harbor had been a traumatic experience, and it had left a deep fear of surprise attack from the air.

Korean War Threats and Massive Retaliation

When President Harry S Truman and his closest advisers decided to drop atomic bombs on Japanese cities, they anticipated that Japan's leaders would be shocked into surrendering. This assessment was correct. Truman and his advisers also expected that Stalin would be awe-struck by the bomb, and that he would become much more compliant on issues that were beginning to divide the two big powers. Here they were only half right. Stalin was indeed much impressed by the bomb, and he immediately ordered a speed-up of the Soviet atomic program. But in negotiations he dug in his heels on the issues that mattered most to Russia.

Stalin absolutely refused to compromise on the future of Poland, an especially sensitive matter because World War II had begun as an effort to defend Poland's freedom against outside aggression. Stalin withdrew his forces from Iran, perhaps under pressure of a direct nuclear threat from Truman,[1] and he withheld support from communist guerrillas in Greece and Turkey. But such moves were far overshadowed, in Western eyes, by the communist takeovers he encouraged in Hungary and Czechoslovakia. After the communist takeover in Hungary, Truman arranged aid for the beleaguered non-communist governments in Turkey and Greece, telling Congress that it should be U.S. policy "to support free peoples who are resisting attempted subjugation by armed minorities or outside pressures." This "Truman doctrine" marked the beginning of a global U.S. effort to "contain" what was seen as an expansionist Soviet state. In the wake of the Czechoslovakian takeover, Truman got Congress to adopt the Marshall Plan, a program of economic aid for Western Europe, which was designed to improve social conditions and undercut the position of the communist parties.

The conflict between the United States and the Soviet Union came to a head in mid-1948 when the Soviet Union blockaded Berlin after the Western powers took steps toward consolidating their occupied zones in Germany. The United States deployed its nuclear forces three times in the first half of 1948 to convey a signal to Russia that it would stop at nothing to defend Berlin.[2] The first Berlin crisis ended with Russia backing down, but Germany was split, and military forces representing two hostile blocs soon were facing each other along the East-West divide.

Despite the precipitous deterioration in U.S.-Soviet relations, neither of the superpowers was eager so soon after a major war to pour national resources into the military's coffers. Both countries substantially demobilized their armies, and both tried to keep defense spending to a minimum. Two years after the war ended there were just 13 atomic bombs (A-bombs) in the U.S. arsenal, and a year later, during the Berlin blockade, 50.[3]

In 1949-1950, after Russia had tested its first A-bomb and communist forces had won the civil war in China, the arms race began to gather speed. In the United States pressure mounted sharply for much tougher anti-communist policies; Sen. Joseph McCarthy, R-Wis., used findings about Soviet atomic espionage to create the impression that the U.S. government had fallen under the influence of communist agents.

In March 1950, six months after the Soviet A-bomb test, Truman ordered a crash program to develop hydrogen bombs. In April the president's National Security Council prepared an important policy document, NSC-68, which called for a rapid buildup of U.S. forces. Two months later communist North Korea attacked South Korea, and Truman immediately dispatched U.S. troops to repulse the North Korean troops. Rep. Lloyd Bentsen, D-Texas, called for the use of atomic weapons in Korea.

On November 26 after U.S. forces had driven the aggressors back and crossed the line into North Korea with the intention of reuniting the country, China intervened on the side of the North. Four days later Truman implied at a press conference that he might use nuclear weapons against China. This threat alarmed the Europeans and England's foreign minister, Clement Atlee, flew to Washington to be reassured that Truman would not actually order the atomic bombing of China.

Truman's policies in Korea were extremely unpopular at home, and during the first months of 1951 a "great debate" erupted in Congress over his conduct of foreign policy. The debate opened with a stinging attack on Truman by Sen. Robert A. Taft, R-Ohio, the leader of Republicans who were wary about foreign involvements. "Without authority he involved us in the Korean War. Without authority he apparently is now adopting a similar policy in Europe," Taft said, referring to Truman's request for added U.S. troops for Europe.[4] Gen. Dwight D. Eisenhower, commander of NATO, returned from Europe to persuade Congress to approve the troop request. In 1952 Eisenhower was elected president, having promised to bring the Korean War to a speedy conclusion.

Eisenhower Deploys 'Tactical Weapons'

Eisenhower's threat to use nuclear weapons, which was conveyed through diplomatic channels to the Chinese, may have been the decisive element in bringing about a prompt end to the Korean conflict.[5] From the beginning, the Eisenhower administration's policy was to emphasize U.S. nuclear superiority in diplomacy to save on manpower and money devoted to conventional defenses. Eisenhower was a fiscal conservative and as a midwestern Republican whose closest political ties were with middle-American big businessmen such as Charles Wilson, his secretary of defense, and George M. Humphrey, his treasury secretary, he sincerely believed that high military expenditures drove up budget deficits and caused long-term damage to the economy. During his first year in office Eisenhower cut the defense budget to $36 billion from $42 billion; the outgoing Democratic administration had planned to increase defense expenditures by $7-$9 billion.[6]

In October 1953 Eisenhower approved NSC-162/2, which called for the creation of contingency plans for using nuclear weapons in limited wars. Consistent with this policy, which was called the "New Look," the United States began the deployment overseas of "tactical nuclear weapons"— weapons intended for battlefield use — even though they were indistinguishable, initially, from other atomic bombs in the U.S. arsenal. Military exercises in Europe were based on the theoretical assumption that nuclear weapons were used, and in the United States soldiers conducted maneuvers close to atomic explosions so that they could be tested and trained for real-life nuclear combat.

In January 1954 Secretary of State John Foster Dulles made it known that the U.S. response to Soviet aggression would be "massive retaliation," in other words, an all-out nuclear attack. In the event war broke out, the U.S. Strategic Air Command's goal would be to leave the Soviet Union "a smoking, radiating ruin at the end of two hours," according to a Navy memorandum based on an Air Force briefing in March 1954.[7] The plan was to attack the entire range of targets in Russia simultaneously, including both cities and military installations. After the United States began to send U-2 reconnaissance planes over the Soviet Union in 1956, planning for attacks on counterforce targets was greatly facilitated, and by 1959 U.S. strategists had identified and analyzed some 20,000 suitable military installations in Russia.[8]

To the Missile Crisis

Following the death of Stalin in 1953, a leadership struggle ensued in Russia from which Nikita Khrushchev finally emerged as victor. At the 20th Communist Party Congress in 1956 Khrushchev denounced Stalin for committing blunders and crimes. At the same time he pronounced that war with the capitalist countries was not "fatalistically inevitable" — a major departure from traditional communist doctrine.

Few people were inclined to take Khrushchev's statements at face value, partly because he had risen to the top as one of Stalin's most trusted aides, and partly because his actions and words often seemed to belie his professions of benign intent. The Soviet Union's harsh suppression of the Hungarian uprising in November 1956 inflamed anti-communist sentiment. At the same time Khrushchev made a well-publicized threat to use nuclear weapons against England when it launched an attack on Egypt, along with France and Israel, after President Gamal Abdel Nasser of Egypt nationalized the Suez Canal. Because the Soviets were going to great lengths to conceal how weak their nuclear forces really were, Khrushchev's boasts often were given more credence in the West than they deserved.[9]

The Eisenhower administration's determination to gather intelligence on Soviet forces and Khrushchev's desire to conceal his hand were major factors in the collapse of the first serious effort at arms control in the late 1950s. In 1957 Khrushchev issued a call for a test ban, and the following year the Soviet Union announced a unilateral halt to testing. The United States joined in the moratorium the same year, and in 1958 scientific representatives from the two superpowers met in Geneva to lay the technical foundation for verification of a test ban.[10] There continued to be some chance that the two countries might negotiate a ban even after the moratorium's one year term expired, but in May 1960 all hopes were dashed when the Soviet Union shot down a U-2 reconnaissance plane over its territory. Eisenhower initially denied that he was sending spy planes over Russia but then admitted it, and Khrushchev used the occasion to break up a Paris summit meeting and issue harsh denunciations of the United States. Eisenhower's reversal on the U-2 occurred, according to Townsend Hoopes in *The Devil and John Foster Dulles*, in no small part because of "Republican vulnerability to the Democratic charge that Eisenhower not only lacked control of the government, but was even unaware of espionage activities that could wreck his stated policy." [11]

By the time of the U-2 incident Eisenhower had come under heavy criticism from a coalition of Democrats and Air Force officers for allegedly failing to counter the Soviet threat with enough foresight and energy. After Sputnik many Americans sincerely believed that the Russians were far ahead in the race to develop intercontinental missiles, and while it later became clear that this was not the case, intelligence from the U-2 flights was not sufficiently comprehensive in the late 1950s to refute charges made by alarmists. In the 1960 presidential campaign the Democratic candidate, John F. Kennedy, said that a "missile gap" had developed under Eisenhower, and he promised to take vigorous measures to close it if elected president.

Kennedy and Khrushchev adopted confrontational postures toward each other from the beginning, and it did not take long for relations between the two leaders to reach a critical state. At their first meeting, which took place in Vienna on June 4, 1961, Kennedy was shocked by Khrushchev's belligerent attitude, and he came away convinced that the Soviet leader considered him a weakling. Just two months before this meeting Kennedy had suffered a humiliating personal defeat when forces sponsored by the CIA were repulsed in an attempted invasion of communist Cuba. On June 21, less than a month after the meeting in Vienna, Khrushchev issued an ultimatum on the status of East Germany that seemed to threaten the freedom of West Berlin.

On July 25, after tension had mounted for a month over Berlin, Kennedy delivered a speech to the nation in which he stressed his determination to defend the city. In addition to calling for increases in defense spending and mobilization of reserves, Kennedy outlined a massive civil defense program that was "designed to signal to the Russians his willingness to run the risk of nuclear war over Berlin," as the historian Robert A. Divine has put it.[12] On August 13 work began on the notorious Berlin Wall, and on August 30 the Soviet Union resumed with atmospheric testing of nuclear weapons for the first time in three years. The United States followed suit the next month.

Andrei Sakharov, who helped develop the Soviet Union's hydrogen bomb (H-bomb), has provided one account of Russia's decision to resume testing. When Sakharov learned that a new test series was to be prepared to bolster the Soviet position on the German question, he says he wrote a note to Khrushchev objecting to the decision on the ground that it "would lead to a new round in the armaments race." At a meeting with scientists, according to Sakharov's account, Khrushchev delivered some off-the-cuff remarks rejecting Sakharov's advice. Khrushchev "more or less said the following," Sakharov reported. "Sakharov is a good scientist. But leave it to us who are specialists in this tricky business to make foreign policy.... We can't say aloud that we are carrying out our policy from a position of strength, but that's the way it must be. I would be a slob, and not the chairman of the Council of Ministers, if I listened to the likes of Sakharov."

Khrushchev's account of the same incident has quite a different flavor. "I replied," Khrushchev reported he said, 'Comrade Sakharov, believe me, I deeply sympathize with your point of view, but as the man responsible for the security of our country, I have no right to do what you're asking. For me to cancel the tests would be a crime against our state. I'm sure you know what kind of suffering was inflicted on our people during World War II. We can't risk the lives of our people again by giving our adversary a free hand to develop new means of destruction.' " [13]

Kennedy and the Cuban Missile Crisis

On Oct. 16, 1962, following weeks of mounting anxiety about what the Russians might be up to in Cuba, Kennedy received word that they definitely were installing nuclear-armed missiles on the island — 90 miles off the U.S. coast. Kennedy immediately established a 14-man group, the so-called "Executive Committee" or "ExCom," to advise him on actions to take. For 13 anxious days this group met — sometimes in expanded, sometimes in reduced, form — to determine how the U.S. government might get the Soviet missiles out of Cuba without provoking a nuclear war. Just a month and a half earlier, in a meeting between the Soviet ambassador, Anatoly Dobrynin, and the president's brother, Robert Kennedy, Dobrynin had told Kennedy that the Soviet Union would not give a third country such as Cuba the power to involve it in a thermonuclear war. [14]

On October 22, in a televised address to the nation, President Kennedy said that offensive missiles had been discovered in Cuba, despite Russian denials. Kennedy announced that he was putting military units on alert and instituting a naval blockade of Cuba. He said that if any missiles were fired from Cuba at any country in the Western

Hemisphere, the United States would respond with a direct attack on the Soviet Union.

On the morning of October 24, tension peaked at the White House when word arrived that a Soviet submarine had positioned itself between U.S. and Soviet ships. At this point, 24 of the Soviet missiles in Cuba were fully operational. According to Robert Kennedy's account, the president's face turned gray and haggard, he put his hand over his mouth and clenched his fist spasmodically. "For a few fleeting seconds," Kennedy said, "it was almost as though no one else was there.... Inexplicably, I thought of when he was ill and almost died; when he lost his child; when we learned that our oldest brother had been killed; the personal times of strain and hurt. The voices droned on but I didn't seem to hear anything." [15]

Within a half hour news arrived that the Soviet ships headed for Cuba had stopped and were turning back. Secretary of State Dean Rusk said to McGeorge Bundy, the president's special assistant for national security, "We're eyeball to eyeball and I think the other fellow just blinked." [16] Later in the week, after receiving contradictory messages from the Soviet Union about their conditions for withdrawing the missiles, Robert Kennedy met privately with Dobrynin. The next day, on October 28, Khrushchev announced that he would withdraw the missiles from Cuba in exchange for a public pledge from the United States not to invade the island.

At the time of the Cuban missile crisis it generally was assumed in the West that Khrushchev quite simply was trying to tip the balance of nuclear power in his favor so as to extract concessions on Berlin and other possibly contentious issues. In the intervening years, however, it has become clear that Khrushchev's strategic position was far weaker than the public appreciated at the time. Having gotten himself into an exposed position by pursuing an aggressive foreign policy, Khrushchev may in fact have been trying to shore up his defenses.

The Soviet Union had deployed only four of its first intercontinental ballistic missile (ICBM) models, and it was beginning to deploy its second model in 1962. The United States, meanwhile, already was well advanced with a rapid ICBM buildup, and it was deploying medium-range missiles not only in England and Italy but also in Turkey, a country immediately adjacent to the Soviet Union. The United States had a commanding lead over Russia in bombers as well as missiles: B-47s already were being phased out in 1962, but the year before the United

States still had more than 2,000 long-range bombers while the Soviet Union had fewer than 200.[17] On June 16, 1962, Defense Secretary Robert McNamara delivered a speech describing a new strategic doctrine of "controlled response," which called for using nuclear weapons against limited military targets before resorting to all-out counter-city attacks. The speech may have revived Soviet fears of a pre-emptive strike.

Six months after the Soviet missiles were withdrawn from Cuba, in April 1963, the United States withdrew its medium-range missiles from Turkey and Italy. For many years thereafter speculation was rife as to whether Kennedy had struck a secret deal with Khrushchev, as Robert Kennedy had hinted in his memoir on the crisis.[18] In an attempt to counter such speculation, it often was said that the Jupiter missiles removed from Turkey and Italy were obsolete and that Kennedy already had ordered their removal before the missile crisis. In 1980, however, Barton Bernstein of Stanford University published an article in which he showed convincingly that 1) the Jupiter missiles had only become operational in Turkey in mid-1962, months before the Cuban missile crisis; 2) Kennedy had ordered a review of their status but not their removal; 3) at every stage of the 13-day missile crisis, a deal involving the exchange of the Turkish missiles for the Cuban missiles was considered; 4) but that a *public* deal of this kind was repeatedly rejected, mainly because of the adverse effect such a deal would have on U.S. prestige among its European allies.[19]

When Bernstein's article appeared, a great deal of evidence had accumulated suggesting that Robert Kennedy and Dobrynin had made a *private* deal calling for the removal of the Turkish missiles in exchange for the Cuban rockets. If the Soviet Union had insisted on a public agreement, apparently, the Kennedy administration would have been willing to take quite drastic military actions, including a direct assault on the missile launchers in Cuba, even though 24 already were loaded with nuclear weapons.

In October 1982, on the twentieth anniversary of the missile crisis, six of Kennedy's top advisers confirmed that a secret deal was indeed struck. Writing in *Time* magazine, they said that the "private assurance — communicated on the president's instructions by Robert Kennedy to Soviet Ambassador Anatoli Dobrynin on the evening of Oct. 27 — was that the President had determined that once the crisis was resolved, the American missiles then in Turkey would be removed. . . ." This second assurance was kept secret, Kennedy's advisers explained, "because the few who knew about it at the time were in unanimous agreement that any other course would have had explosive and destructive effects on the security of the United States and its allies. If made public in the context of the Soviet proposal to make a 'deal,' the unilateral decision reached by the President would have been misread as an unwilling concession granted in fear at the expense of an ally [Turkey]." [20]

Kennedy's former advisers ended their *Time* piece with a paean to the two men who resolved the crisis without going to war. "We know," they said, "that in this anniversary year John Kennedy would wish us to emphasize the contribution of Khrushchev; the fact that an earlier and less prudent decision by the Soviet leader made the crisis inevitable does not detract from the statesmanship of his change of course."

Even at the time, Khrushchev's willingness to be personally humiliated for the sake of preventing war was widely appreciated, and during the following year relations between the two superpowers warmed considerably. In August 1963 the United States, the Soviet Union, and England concluded a partial test ban treaty, which prohibited atmospheric testing of nuclear weapons. The partial test ban was a useful environmental measure in that it reduced fallout from nuclear tests,* but as an arms control measure it was disappointing. In the years following the conclusion of the treaty, the two superpowers tested more nuclear weapons each year than they had before 1963.

In a number of ways the Cuban missile crisis left the world with a somewhat false sense of security. For one thing, the accords that ended the crisis were complex — partly formal and public, partly informal and secret. Even today, few people are fully aware of what the agreements were, which makes it easy for American and Russian politicians to score points by accusing each other of violating the accords or the spirit of the accords. [21]

Khrushchev's handling of the missile crisis almost certainly was a significant factor in his fall from power in October 1964, when a more militarily minded leadership took power in the Soviet Union. Since 1964 the group that ousted Khrushchev has engaged in a gradual but relentless buildup of Russia's overall military forces — a departure from the policies of Khrushchev, who sought to strengthen the Soviet strategic arsenal but to economize on conventional weapons and

---

* France and China did not sign the partial test ban and continued with atmospheric testing, ignoring protests about fallout.

Origins of the Arms Race
and the Blocs

August 1945	U.S. bombing of Hiroshima and Nagasaki.
March 1946	Winston Churchill, speaking at Fulton, Mo., says "iron curtain" has fallen over Eastern Europe.
January 1947	Communists gain control of Hungarian Government.
March 12, 1945	"Truman doctrine" speech to Congress.
July 1947	George F. Kennan's article on "Sources of Soviet Conduct" appears in *Foreign Affairs:* describes Russia as expansionist and calls for containment policy.
March 1948	Communists take control of Czechoslovakian government; Truman calls for Marshall Plan to aid West European economies.
June 1948	Soviet Union blockades Berlin after the United States, England, and France introduce a new currency in the zones they control in western Germany.
April 1949	NATO treaty takes force.
May 1949	Federal Republic of West Germany formed.
September 1949	Washington announces first Soviet A-bomb test.
December 1949	Washington acknowledges victory of Chinese communists.
Feb. 3, 1950	Klaus Fuchs in London confesses to atomic espionage.
Feb. 9, 1950	Sen. Joseph McCarthy, R-Wis., says he has list of 205 communist spies in U.S. State Department.
April 1950	NSC-68 calls for major buildup of U.S. forces.
June 1950	North Korea attacks South; Russian collaboration suspected.
October 1950	Chinese communists intervene in Korean War.
Winter 1951	"Great Debate" in U.S. Congress over added troops for Europe.
1954	Paris Pacts establish terms for remilitarization of West Germany.
1955	Warsaw Pact founded.

manpower. Among western analysts it almost universally is assumed that Khrushchev's successors made it an overriding goal never again to be humiliated as Krushchev had been over the Cuban missiles.

In the United States the dramatic victory in the missile crisis may have caused Kennedy's advisers to become over-confident and too unwilling to reconsider policies of dubious merit. During the years immediately preceding the missile crisis, it was disclosed in investigations conducted by a Senate committee in 1976, the CIA repeatedly tried to have Castro assassinated.[22] Even though the CIA's campaign against Castro may have been a decisive factor in his willingness to invite the Soviet Union to deploy missiles on the island, the CIA continued with its attempts to "get" him even after the crisis ended.[23] Whether the Kennedy brothers themselves knew of the assassination attempts is a controversial question.[24]

Nixon's Vietnam Alert

During the months just before his assassination, Kennedy was becoming more doubtful about the growing U.S. involvement in South Vietnam where American military "advisers" were trying to shore up a weak and unpopular government in the face of a popular insurgency supported by the communist North. Around this time Kennedy took a number of measures to limit the U.S. military commitment. He rejected a plan for covert operations to be launched against North Vietnam, which the Joint Chiefs of Staff sponsored, and he called for a reduction of one thousand advisers. Two days after Kennedy's assassination, however, plans to pare back the U.S. involvement in Vietnam were changed. As recounted by Richard M. Pious of Columbia University, President Lyndon B. Johnson held a meeting with military advisers, resulting in the restoration of full military and economic aid to South Vietnam. The central objective of Johnson's new policy, which was spelled out in National Security Memorandum 273, "was no longer to assist South Vietnam in its struggle," Pious concluded; "it had now become an American commitment to 'win' the fight against the Communists, an open-ended military commitment." [25]

Between 1963 and 1968 the number of troops Johnson committed to the war increased to more than 500,000 from around 20,000. The conflict rapidly changed from a civil war, in which U.S. advisers helped a client government defend itself against a guerrilla insurgency supported by the North, into a direct confrontation between the United States and

North Vietnam. Beginning in 1965, after Congress authorized Johnson to conduct the war as he saw fit with the "Gulf of Tonkin resolution," the U.S. Air Force began to bomb North Vietnam and vast tracts of the South on a regular basis. There is no evidence that top officials ever seriously considered using nuclear weapons in Vietnam, but they often reminded the public of what they referred to as their "restraint," which seemed to indicate they were not completely excluding measures even more far-reaching than the ones they already had adopted. In the fast growing anti-war movement such statements were received with consternation; in the eyes of the war's critics, U.S. actions already had reached a level of destructiveness that could not be justified in terms of any identifiable national interest.

In 1968 Johnson withdrew from the presidential race when it became apparent that the Democratic Party was deeply divided over his conduct of the war. Vice President Hubert H. Humphrey lost the election to Richard Nixon, who said during the campaign that he had a plan to bring the war to an early conclusion. At the time it was not clear whether he was intending to withdraw quickly from Vietnam, much as de Gaulle — one of Nixon's heroes — had withdrawn France from the controversial Algerian war in the late 1950s, or whether he hoped to break North Vietnam's will with threats of the kind Eisenhower issued to North Korea and China in 1953.

Soon after Nixon took office it became apparent that he was opting for the Eisenhower model. According to his memoirs, Nixon decided in the middle of his first year in office to "go for broke" and try to end the war "either by negotiated agreement or by an increased use of force." [26] On July 15 he sent an ultimatum to North Vietnam saying that unless a breakthrough was achieved in negotiations by November 1, he would find himself "obliged to have recourse to measures of great consequence and force. . . ." In the weeks leading up to November 1 Nixon said he sought to "orchestrate the maximum possible pressure on Hanoi," in the hope of convincing the North Vietnamese that he meant business. According to Nixon's aide, Harry R. "Bob" Haldeman, the president told him he was following the same strategy Eisenhower had pursued when he let the Chinese know he would use nuclear weapons in Korea unless a truce was concluded quickly. "I call it the Madman Theory," Haldeman claims Nixon said. "I want the North Vietnamese to believe I've reached the point where I might do anything to stop the war." [27]

In October 1969 Nixon ordered the Strategic Air Command to place its nuclear-armed B-52 aircraft on alert. This was the first time such an alert had been ordered since the Cuban missile crisis, and it is reported to have lasted 29 days, although the U.S. public knew nothing of it at the time.[28] Nixon later claimed that he already knew by October 14 that his ultimatum had failed. The reason, he said in his memoirs, is that a quarter of a million people were to turn out the next day for the first big "Moratorium" demonstration against the war in Washington. The anti-war movement, he believed, had "destroyed whatever small possibility may still have existed of ending the war in 1969." [29]

During the following years Nixon tried to bring added pressure on the North, mainly by extending the war into Cambodia. First, he ordered secret bombings of North Vietnamese "sanctuaries" just inside the Cambodian border; then an "incursion" of U.S. forces into Cambodia to attack North Vietnamese command centers; finally he sponsored an invasion of the country by the South Vietnamese. These moves did much to bring on a national tragedy of dimensions unparalleled anywhere in the world since 1945,[30] but they did not visibly affect the will of the North Vietnamese, and Nixon's actions fueled anti-war activity in the United States. Growing numbers of people continued to turn out for demonstrations in Washington, and at times during these confrontations Nixon showed alarming signs of mental distress.[31]

Détente and Salt I

One of the great ironies of Nixon's administration was his achievement of a breakthrough in U.S.-Soviet relations at the very time when U.S. troops were fighting in Vietnam against forces armed by Russia. A complicated mixture of idealistic and pragmatic considerations probably figured in his decision to negotiate a strategic arms treaty with the Soviet Union. One significant factor certainly was the need to economize on non-Vietnam-related defense expenditures. There was especially grave concern that efforts by each side to develop and deploy anti-ballistic missile (ABM) systems would unleash an expensive new round in the arms race, which would be dangerous or futile. If workable defenses against ballistic missiles could be built, the first superpower to develop an ABM system might be tempted to launch a strike against the other in hopes of winning a pre-emptive nuclear war. More likely, in the opinion of many weapons analysts, it would be impossible to develop a really effective ABM, and any effort to do so would be wasteful and deceptive.

As early as 1966, Defense Secretary McNamara began to press for deferral of work on an ABM system until possibilities for arms control had been explored with the Soviet Union. At a meeting between Johnson and Soviet leader Aleksei Kosygin in 1967 at Glassboro, N.J., McNamara urged the Soviet leader to consider an ABM ban, but without apparent success. The Russians soon reconsidered their position, however, and signaled their willingness to discuss an ABM treaty provided offensive systems also were limited.

On June 27, 1968, Soviet Foreign Minister Andrei A. Gromyko informed the Supreme Soviet — the Soviet Union's legislative body — of the Kremlin's readiness to discuss "mutual limitation and subsequent reduction of strategic means of delivery of nuclear weapons, both offensive and defensive, including antiballistic missiles." Five days later Johnson announced the two governments would open strategic arms limitation talks.

President Nixon took office in 1969 with grave reservations about his predecessor's arms control efforts and was predisposed to proceed with deployment of a U.S. ABM system. But from a technical point of view an ABM looked increasingly unattractive, and, with Vietnam War costs rising, Congress had launched an economy drive and was beginning for the first time in a decade to scrutinize new weapons requests carefully. In August 1969 Vice President Spiro T. Agnew had to cast a tie-breaking vote in the Senate to prevent Congress from cutting ABM funding.

In these circumstances Nixon authorized strategic arms negotiations. His national security adviser, Henry A. Kissinger, proceeded to gather control over the negotiations and gradually overcame opposition to an agreement within the U.S. government. In May 1972 President Nixon traveled to Moscow to sign, amid much pomp and splendor, the ABM Treaty and the Interim agreement on Offensive Arms — the SALT I accords.

Ratification of the ABM treaty occurred with remarkable ease. The Senate approved the treaty by a decisive vote of 88 to 2 on Aug. 3, 1972, just three months after it had been concluded. The Interim Agreement in principle did not require congressional authorization because it was an executive agreement and not a treaty, but Nixon submitted it to both the House and Senate for approval.

SALT I Agreements

President Richard Nixon and Chairman Leonid Brezhnev signed two strategic arms limitation pacts in Moscow on May 26, 1972. One was the Treaty on the Limitation of Anti-Ballistic Missile Systems (ABMs), and the other was technically an executive agreement placing a numerical freeze on U.S. and Soviet missile launchers for five years at roughly the existing levels.

Although both agreements are sometimes referred to as SALT I, current tendency is to speak of SALT I in terms of the second agreement — on offensive weaponry. The other is usually called the ABM treaty.

That treaty limited each side to one ABM site for defense of Moscow and of Washington and to one site for the defense of an intercontinental ballistic missile (ICBM) "Field," or facility — in the United States, at Grand Forks, N.D. No Washington-area ABM site was ever put into operation, however, and the Grand Forks installation was shut down soon after the treaty was concluded.

The agreement on offensive weaponry limited ICBMs to the number under construction or deployed as of July 1, 1972 — about 1,618 for the Soviet Union and 1,054 for the United States — and froze deployment of nuclear submarines at existing levels — 62 Russian and 44 American.

The Interim Agreement proved to be more controversial than the treaty, primarily because it allowed the Soviet Union to keep a larger number of missiles than the United States had deployed. Several senators proposed modifications to the agreement. The most important was an amendment sponsored by Sen. Henry M. Jackson, D-Wash., which the Nixon administration accepted, stating that any future permanent strategic arms treaty should "not limit the United States to levels of intercontinental strategic forces inferior to" those of the Soviet Union, but rather should be based on "the principle of equality."

Jackson's amendment implied that the SALT I accords were *not* based on the principle of equality, even though they left the United States with a much larger number of nuclear warheads on its missiles than the

Soviet Union had. Having persuaded Congress to adopt this dubious interpretation of SALT I, Jackson would use his amendment as a basis for criticizing proposed SALT II treaties throughout the 1970s. Meanwhile, as a condition for his support for the SALT I agreement, Jackson got Nixon to purge the U.S. Arms Control and Disarmament Agency (ACDA) of people considered too enthusiastic about arms control. In December 1972 Nixon's Office of Management and Budget ordered ACDA to cut its budget by a third and to fire specified officials. The staff of the agency, which originally was smaller than the Pentagon's public relations office, fell from 219 to 162.[32]

Many of the Americans who were involved in negotiating the SALT accords hoped that they would inaugurate a new era in which the superpowers could conduct their relations more openly and in an atmosphere of trust. Russian leaders apparently thought the accords would be the beginning of an age in which the United States and the Soviet Union would work more closely together, as equals, to resolve issues. The "Basic Principles" of the accords — a document adopted mainly at the urging of the Soviet Union — said that the superpowers should conduct their relations on the basis of equality, reciprocity, mutual recognition, and benefit.[33]

The more extravagant hopes on both sides were dashed with the outbreak of a war in the Middle East in October 1973. Some Americans felt that the U.S. government was betrayed because the Soviet Union had not warned them of Egypt's surprise attack, even though such a warning would have undermined the Soviet Union's reputation as a protector of Arab states. The Russians were disappointed because Nixon's reputation was to refuse to deal as equals with them in resolving the crisis, even though it long had been an aim of U.S. policy to minimize Soviet influence in the Middle East.

1973 Nuclear Alarms

When Egypt and Syria attacked Israel in October 1973 the Watergate affair had reached a serious stage, creating uncertainty in the United States and abroad as to whether the Nixon administration would be able to respond effectively to a sudden emergency. The war was a complete surprise to top administration officials, who had misread developments in the Middle East. From the beginning, their main concern was that the Soviet Union would regard U.S. leadership as weak and would try to exploit the situation.

Following the 1967 Middle East War, in which Israel crushed Egypt's forces in just six days, the Soviet Union made a massive commitment to the rearmament of Egypt. But in 1971 Anwar Sadat, Egypt's president, expelled Soviet advisers. U.S. officials tended to interpret Sadat's move as a friendly gesture toward the West, when in fact it amounted to a rejection of the Soviet Union's advice to refrain from going to war against Israel. After Sadat launched the war against Israel on Oct. 6, 1973, the initial Soviet reaction was ambiguous; some scholars have described it as "restrained," while others have seen it as an opportunistic attempt to make the most of a troubled situation — as long as Egypt remained ahead.

By the end of the first week of the war, when Egypt still was mounting a strong offensive against Israel in the Sinai, the Soviet Union began to airlift supplies to Egypt and Syria. On October 13 a U.S. airlift to Israel began. Two days later, without notifying the U.S. government of its intentions, Israel launched a daring counterattack on Egypt's side of the Suez. Within days the tank columns commanded by Gen. Ariel Sharon were posing a mortal threat to Egyptian forces.

On October 18 Soviet ambassador Dobrynin proposed to Secretary of State Kissinger that a cease-fire be put in place, to be guaranteed by the Soviet Union and the United States. Kissinger rejected the proposal on the ground that Israel would not agree to it, but late the next evening he flew to Moscow at Brezhnev's request to discuss the situation. On Monday October 22, after Kissinger had reached an agreement with the Russians, the U.N. Security Council adopted a cease-fire resolution. But this resolution and a second one did not hold; meanwhile, Israeli forces had surrounded Egypt's Third Army, threatening to completely destroy it.

Now that the tide had turned, U.S. intelligence reports indicated on October 24 that Soviet airborne divisions were being placed on alert, and that evening Dobrynin gave Kissinger an urgent personal message for Nixon from Brezhnev. The Soviet leader proposed that the two superpowers should "urgently dispatch" contingents to the Middle East to implement a cease-fire. "I will say it straight," Brezhnev said, "that if you find it impossible to act together with us in this matter, we should be faced with the necessity urgently to consider the question of taking appropriate steps unilaterally." [34] In a response issued that same evening Nixon vetoed the joint dispatch of peace-keeping forces and proposed, instead, a force composed of troops from non-nuclear countries. Later that same night the White House ordered a worldwide nuclear alert. On

October 25, the next day, Kissinger indignantly denied that considerations connected with Watergate played any part in the decision to order
the alert. He said the alert had been necessary because of the very harsh
note the president had received from the Soviet Union.

At the time of the 1973 alert, suspicion was widespread in the
United States that Nixon and/or his advisers had ordered the move to
heighten the crisis atmosphere and deflect attention from Watergate.
Kissinger heatedly denied such charges at the time, and they never have
been conclusively substantiated. It remains unclear, however, exactly how
the alert was ordered and why. There is no doubt, in any event, that
Nixon and some of his advisers were trying to use the crisis to shake
themselves free of the Watergate investigation.

On October 10, four days after the Egypt-Israeli war had begun,
Vice President Agnew was forced to resign because criminal charges
were about to be brought against him. With his departure a major
obstacle to the impeachment of Nixon was removed.[35] Five days later, on
October 15, Nixon's chief of staff, Gen. Alexander Haig, summoned
Attorney General Elliot Richardson, ostensibly for a briefing on the
Middle East war and U.S.-Soviet relations. At the end of the briefing, in
which Haig stressed how critical the world situation was, he told
Richardson that Watergate special prosecutor Archibald Cox would have
to be fired because his investigation was "causing an intolerable diversion
of the president's time and energy from far more important matters."[36]
Richardson had served in two Cabinet positions for Nixon before taking
over as attorney general earlier in the year when Richard Kleindienst was
forced to resign under threat of criminal indictment. Apparently, Haig
thought Richardson would continue to be a team player, but Richardson
said he would resign if Cox was fired.

Five days after Haig's meeting with Richardson, the day Kissinger
flew to Moscow for urgent consultations with Brezhnev, Nixon fired
Cox in what came to be known as the "Saturday Night Massacre." When
Richardson resigned, Nixon personally tried to get him to defer the step
until the Middle East crisis had ended. When Richardson refused,
Nixon's parting words were, "Brezhnev would never understand if I let
Cox defy my instructions." [37]

Conclusions: Probability of Nuclear Attacks

Apart from Nixon's alerts of 1969 and 1973, neither of the
superpowers has issued an overt nuclear threat since the Cuban missile

crisis.[38] The Nixon alerts came under heavy criticism, partly because of doubts about Nixon's judgment and character, but mainly because of a growing conviction that nuclear threats were an inappropriate instrument of diplomacy under circumstances in which either of the superpowers could devastate the other.

During the period when the United States was decisively superior to the Soviet Union, in contrast, the U.S. government was not shy about using its nuclear arsenal for political ends. The U.S. government bombed Hiroshima and Nagasaki to hasten Japan's surrender and to improve the U.S. bargaining position vis-a-vis the Soviet Union; between 1945 and April 1963, when the Jupiter missiles were removed from Turkey, the government changed the alert status or deployment of its nuclear forces to send a threatening signal at least 17 times; and presidents have made direct nuclear threats, publicly or privately, on several occasions — at least twice in connection with the Korean War, and once each over Berlin, Cuba, and possibly Iran.

In the same period, apart from its attempt to deploy nuclear missiles in Cuba, the Soviet Union issued a specific nuclear threat only once, and it was directed against England, a vastly inferior nuclear power, not the United States. Khrushchev made the threat in 1956 when British and French forces joined with Israel in an attack on Egypt. Despite his restraint in issuing nuclear threats, Khrushchev often talked about the Soviet Union's atomic might in a way that was bound to alarm people in the West. In September 1961 he casually told a reporter for the *New York Times* that England, France, and Italy would refuse to join the United States in military action in a showdown over Berlin. He said that such countries were hostages to the Soviet Union and a guarantee against war. During the Berlin crisis Khrushchev is reported to have told the British ambassador that Western Europe was at his mercy: he said six hydrogen bombs would annihilate the British Isles and nine others would take care of France.

Notes

1. Sen. Henry M. Jackson, D-Wash., later said Truman issued a nuclear threat to get Stalin to withdraw his forces from Iran. See *Time*, Jan. 28, 1980.

2. See Barry M. Blechman and Stephen S. Kaplan, *Force Without War: U.S. Armed Forces as a Political Instrument* (Washington, D.C.: The Brookings Institution, 1978), 48.

3. Michel Tatu, "Le Vrai Départ de la Course a la Bombe," *Le Monde*, Oct. 16-17, 1983, 2, and Thomas B. Cochran et al., *Nuclear Weapons Data Book*, I (Cambridge, Mass.: Ballinger Publishing Co., 1983), 15.

4. *Congressional Quarterly Almanac, 1951* (Washington, D.C.: Congressional Quarterly, 1951), 221.

5. See Dwight D. Eisenhower, *Mandate for Change, Vol. I* (Garden City, N.J.; Doubleday & Co., 1963), 178-181.

6. See Townsend Hoopes, *The Devil and John Foster Dulles* (Boston: Little, Brown & Co., 1973), 193.

7. David Alan Rosenberg, "A Smoking Radiating Ruin at the End of Two Houses," *International Security* (Winter 1981-1982).

8. Ibid.

9. For example, at the Moscow May Day parade in 1955 the Soviets had their planes circle back and fly over the city repeatedly to create the impression that their air forces were much larger than they really were. For U.S. intelligence in this period and the origin of the "bomber gap" and "missile gap" myths, see Lawrence Freedman, *U.S. Intelligence and the Soviet Strategic Threat* (Boulder, Colo.: Westview Press, 1977) and John Prados, *The Soviet Estimate: U.S. Intelligence Analysis and Russian Military Strength* (New York: Dial Press, 1982).

10. See Robert Gilpin, *American Scientists and Nuclear Weapons Policy* (Princeton, N.J.: Princeton University Press, 1962).

11. Hoopes, *The Devil*, 501.

12. Robert A. Divine, *Since 1945: Politics and Diplomacy in Recent American History* (New York: John Wiley & Sons, 1979), 117.

13. Both Sakharov and Khrushchev are quoted in Herbert F. York, "Sakharov and the Nuclear Test Ban," *The Bulletin of the Atomic Scientists* (November 1981): 35-36.

14. Herbert S. Dinerstein, *The Making of a Missile Crisis, October 1962* (Baltimore: The Johns Hopkins University Press, 1976), 181.

15. Robert F. Kennedy, *Thirteen Days* (New York: W. W. Norton & Co., 1969), 69-70.

16. Quoted in *The Saturday Evening Post*, Dec. 8, 1962.

17. David Holloway, *The Soviet Union and the Arms Race* (New Haven: Yale University Press, 1983), 43, and David C. Morrison, "Air-breathing Nuclear Delivery Systems," *The Bulletin of the Atomic Scientists* (February 1983): 33.

18. Kennedy, *Thirteen Days*, 107-109.

19. Barton Bernstein, "Cuban Missile Crisis: Trading the Jupiters in Turkey?" *Political Science Quarterly* (Spring 1980): 97-125.

20. Dean Rusk, Robert McNamara, George W. Ball, Roswell L. Gilpatric, Theodore Sorensen, and McGeorge Bundy, "The Lessons of the Cuban Missile Crisis," *Time*, Sept. 27, 1982, 86.

21. In the public accords, the United States promised not to invade Cuba and the Soviet Union promised not to introduce "offensive weapons" into Cuba, which is generally taken to mean "nuclear weapons." See McGeorge Bundy, "History Warns...," *The New York Times*, Jan. 6, 1984. In 1979, for example, prospects for Senate ratification of SALT II received a blow when Sen. Frank Church, D-Idaho, charged that the Soviet Union had introduced a "combat brigade" into Cuba.

22. Senate Select Committee to Study Government Operations with Respect to Intelligence Activities, *Interim Report-Alleged Assassination Plots Involving Foreign Leaders*, 93d Cong., 2d sess., 1975.

23. See Bill Moyers' Journal, "The CIA's Secret Army," WNET/13, New York, Feb. 13, 1981.

24. See Thomas Powers, *The Man Who Kept the Secrets* (New York: Knopf, 1979) and letter to *The Atlantic* from Frank Mankiewicz and Alan Walinsky, December 1979. See also Arthur M. Schlesinger, *Robert F. Kennedy and His Times* (Boston: Houghton Mifflin, 1978), 485-498.

25. Richard M. Pious, *The American Presidency* (New York: Basic Books, 1979), 383.

26. Richard M. Nixon, *The Memoirs of Richard Nixon* (New York: Grosset & Dunlap), 393.

27. Harry R. "Bob" Haldeman, *The Ends of Power* (Boston: G.K. Hall & Co., 1978).

28. Seymour M. Hersh, *The Price of Power* (New York: Summit Books, 1983), 124-125.

29. Nixon, *Memoirs*, 393-414.

30. See William Shawcross, *Sideshow: Kissinger and Nixon and the Destruction of Cambodia* (New York: Simon & Schuster, 1979).

31. For example, the night of May 8-9, 1970, the weekend of a major anti-war demonstration in Washington, Nixon got up in the middle of the night, visited with demonstrators at the Lincoln Memorial and finally had himself driven to the Capitol with his personal valet, Manola Sanchez. After having his party admitted to the empty House of Representatives early Sunday morning, Nixon sat in his old seat and had Sanchez deliver a speech from the podium. See William Safire, *Before the Fall* (Garden City, N.J.: Doubleday & Co., 1975).

32. See Duncan L. Clarke, *Politics of Arms Control* (New York: The Free Press, 1979), 52-53.

33. See Holloway, *The Soviet Union*, 89.
34. See Tad Szulc, *The Illusion of Peace* (New York: The Viking Press, 1978), 745.
35. Among most people who were inclined to support the impeachment, the succession of Agnew would have been unacceptable.
36. Tad Szulc, *The Illusion*, 743. In this particular incident, this is what Richardson told Szulc.
37. Ibid.
38. See Blechman and Kaplan, *Force Without War*, 48 and Stephen S. Kaplan, *Diplomacy of Power*, (Washington, D.C.: The Brookings Institution), 54. In January 1968, however, the United States deployed its nuclear forces in a threatening manner after North Korea seized the U.S. spy ship Pueblo.

President Jimmy Carter and Soviet Leader Leonid Brezhnev exchange documents after completing the SALT II signing ceremony in Vienna, Austria, on June 19, 1979.

Chapter 10

THE AMERICAN STRUGGLE
OVER STRATEGIC ARMS CONTROL

During the years following President Richard Nixon's fall from power, the question of whether to continue with his policies of détente emerged as a key issue in U.S. politics. Members of the foreign policy elite — diplomats and academic specialists, lawyers, bankers, and business executives — tended to look upon the politicization of U.S. arms control policy with distaste. They often argued that U.S. foreign policy should be based on "objective" national interest — not the vagaries of domestic politics. They remembered with nostalgia the 1950s and 1960s as a kind of golden age when U.S. foreign policy supposedly was based on bipartisan consensus.

Actually, even before the Vietnam War divided the American people, U.S. foreign and military policies often had aroused intense political controversy. Presidents frequently felt obliged to go to considerable lengths to win bipartisan support for military commitments, and much of the time profound differences of opinion on key foreign policy questions emerged within both the Democratic and the Republican parties.

The Politics of Defense: 1945-1974

During the first years of the Cold War, Harry S Truman faced a challenge from former vice president Henry A. Wallace, who had once been considered the front-runner to succeed Franklin D. Roosevelt as president. Wallace accused Truman of not trying hard enough to seek accommodation with the Soviet Union, for pursuing policies that served the interests of the Wall Street business establishment, and for giving indiscriminate support to regimes so long as they were anti-communist. Wallace, a native of Iowa, was heir to the populist movements that swept the Plains States at the end of the 19th century. He first served in government as Roosevelt's secretary of agriculture.

Wallace's "progressivism" met with a sympathetic response almost exclusively among people in the Middle West and among liberal Democrats in New York and Los Angeles. Truman found support for his foreign policy, on the other hand, in a majority of the big northeastern labor unions, among most southern Democrats (despite their growing unhappiness with his civil rights policies), and in the internationalist wing of the Republican Party. The Republican isolationists led by Sen. Robert A. Taft of Ohio remained a force to be contended with. But Sen. Arthur H. Vandenberg, R-Mich., the leader of the internationalists, had emerged as the more influential figure in Washington.

In 1948 when Truman sought congressional funding for the Marshall Plan, Vandenberg advised him to "scare hell" out of the U.S. public by stressing the communist threat to Western Europe. This tactic worked well, and when the Marshall Plan came to a vote, as *Congressional Quarterly* pointed out at the time, "not a single senator from New England or the Pacific Coast was in favor of cutting funds for Europe." Southern Democrats supported the Marshall Plan, and in fact the only real opposition came from representatives of the Middle West — the traditional home of isolationist and populist sentiment. "Out of 77 Republicans against Interim Aid [for Europe]," according to *Congressional Quarterly*, "60 came from the center of the country." [1]

Henry Wallace's candidacy was given little chance in the 1948 election, but Truman was not expected to beat Thomas Dewey, the Republicans' internationalist nominee for president. Even after Truman's upset victory, his authority remained weak in his own party. When Truman committed troops to Korea in 1950, Sen. Richard B. Russell, D-Ga., set up a special Defense Preparedness Committee and appointed his protegé, Sen. Lyndon B. Johnson, D-Texas, to chair the new panel. Johnson denounced Truman for failing to mobilize adequately for war, and he called for round-the-clock industrial production with special emphasis on aircraft development.

After Gen. Dwight D. Eisenhower became president, having been recruited by the internationalist Republicans to head off Taft and the neo-isolationists, there was a period in which little partisan controversy developed over foreign and military policy. As the 1956 election approached, Democrats revealed a deep indecisiveness about how to take issue with Eisenhower's foreign and military policies. In October 1956, when the Democratic presidential candidate, Adlai E. Stevenson of Illinois, got Sen. Stuart Symington, D-Mo., to appear with him in a televised appeal for a nuclear test ban, Symington used the occasion mainly to denounce Eisenhower for failing to put greater emphasis on the development of long-range missiles.[2]

Stevenson's support for a test ban did not galvanize the U.S. public. Yet Symington's "missile gap" took off with Sputnik in October 1957, and it may have helped to carry John F. Kennedy into the White House in 1960. After Sputnik, Johnson, who by then had become Senate majority leader, established himself as the chairman of the new bipartisan Preparedness Subcommittee and a new Committee on Aeronautical and Space Sciences, where he was well placed to criticize Eisenhower for allegedly allowing the nation's defenses to fall into disrepair. Johnson's colleague, Sen. Henry M. Jackson, D-Wash., opened hearings on the foreign policy apparatus in which certain experts took the president to task for being tired and unimaginative in the field of foreign policy.

In the 1960 presidential election Kennedy and his vice presidential running mate Johnson ran on a platform that said the first task of a Democratic administration would be to "restore our national strength" and to "recast our military capacity in order to provide forces and weapons of a diversity, balance and mobility sufficient in quantity and quality to deter both limited and general aggression."[3] As it happened, the Republicans adopted almost identical language on military preparedness, conceding in effect that they shared the view propounded by Symington, Johnson, Jackson, and Kennedy. Judging from this apparent bipartisan unanimity, few people would have guessed in 1960 that both parties soon would be split deeply over military policy and that the divisions later would contribute greatly to the political demise of two powerful presidents — Lyndon Johnson and Richard Nixon.

When public opposition to the Vietnam War came to a head in the 1968 and 1972 elections the leaders of the anti-war forces came from the center of the country, the traditional seat of populism, Progressivism, and isolationist sentiment. Eugene McCarthy, a 1968 Democratic presidential hopeful, won early primary victories that helped persuade Johnson to withdraw from the 1968 race. McCarthy was a senator from Minnesota. Sen. George McGovern, the Democratic challenger to Nixon in 1972, represented South Dakota.

Nixon owed his election in 1968 in part to the inability of the Democratic candidate Hubert H. Humphrey to disassociate himself from Johnson's Vietnam policies and in part to the defection of southern states over civil rights issues. While Nixon easily beat McGovern four years later, his staff's unfair and illegal campaign practices led to his ultimate resignation during the Watergate scandal.

'Soviet Adventurism' and 'Vietnam Syndrome'

For six years after Nixon's forced resignation in August 1974 presidential leadership in the United States was relatively weak. Gerald R. Ford and Jimmy Carter sought to pursue centrist policies, but adverse economic circumstances were polarizing Americans and divisions that emerged over the Vietnam War continued to deepen. Veterans of the anti-war movement wanted to see a radical change in U.S. foreign policy: a much stronger emphasis on arms control and disarmament combined with more attention to humanitarian concerns. Hardliners who had supported the war argued that U.S. foreign policy was being crippled by the "Vietnam syndrome," that is to say, a debilitating reluctance to make use of armed force to defend and advance U.S. interests abroad, even when confronted with Soviet military action.

When Nixon left office the balance of power in Congress — especially in the House — had shifted in favor of a faction that tended to be critical of U.S. military involvements. On Nov. 7, 1973, just after the Middle East war had ended, Congress overrode Nixon's veto and passed the War Powers Resolution, which limited the authority of the president to commit troops overseas. At the end of 1975, when the Soviet Union and Cuba were helping a Marxist regime in Angola consolidate its position after defeating Portugal in a colonial war, the Senate and House passed amendments prohibiting the CIA from providing any further aid to an opposing faction.[4]

In November 1974 Ford and Soviet leader Leonid Brezhnev met at Vladivostok in the Soviet Union and agreed on overall guidelines for a second strategic arms agreement. The guidelines set a ceiling on strategic delivery vehicles — missiles and bombers — of 2,400 and a sub-limit of 1,320 on missiles that could be equipped with multiple warheads. By this time, however, the American public was becoming more receptive to the argument that U.S. meekness was emboldening the Soviet Union, and critics of détente were finding ways to block any further easing of relations.

In 1975 the Soviet Union abrogated its trade agreement with the United States after Senator Jackson pushed an amendment through Congress that made extension of most-favored-nation status to the Soviet Union contingent on its adoption of more liberal emigration policies for Jews. The dispute over trade and emigration, together with disagreements about what weapons should be included in a new arms control agreement, impeded negotiation of a SALT II agreement. As the 1976 election approached, President Ford became increasingly fearful of attacks from hard-line Republicans led by Ronald Reagan. He put the SALT negotiations on hold and declared "détente" a taboo word for administration officials.

Meanwhile the Republican administration was re-evaluating the scope of the Soviet Union's military effort. The size and speed of the Soviet strategic buildup in the late 1960s had taken U.S. intelligence analysts by surprise, and from the beginning of the 1960s up until 1967-1968 the CIA consistently had underestimated the numbers of missiles that Russia would deploy in the years just ahead. In reaction to mounting pressure from organizations such as the Committee on the Present Danger, a group founded in 1976 to draw attention to what it called the "Soviet threat," CIA Director George Bush appointed a special panel of independent experts to prepare an alternative to the CIA's National Intelligence Estimate on Russia. This so-called "Team B" committee concluded that the U.S. intelligence agencies had been regularly underestimating the scope of Soviet military plans. The head of Team B, Richard Pipes, published op-ed pieces in leading newspapers saying that the Soviet Union was intent on unbounded expansion of its global power. Pipes, a professor of Russian history at Harvard University, later became senior adviser on the Soviet Union for President Reagan's National Security Council.

In 1976, his second and final year in office, President Ford decided to increase U.S. defense expenditures by 3 percent annually. Jimmy Carter said during the 1976 presidential campaign that he would cut the defense budget, but after taking office he reversed his position and adopted Ford's defense program. In addition Carter put strong pressure on the NATO allies to raise their annual defense outlays by 3 percent.

Two months after Carter took office Secretary of State Cyrus Vance went to Moscow with a hastily formulated proposal for deep arms cuts, which the Soviet Union rejected out of hand. Foreign Minister Andrei Gromyko charged that the United States was trying to win "unilateral ad-

CIA National Intelligence Estimates
of Soviet Missile Buildup

Year of CIA Prediction	Number of Missiles CIA Predicted Russia Would Build in					
	1967	1968	1969	1970	1971	1972
1961	500					
1962	450	590				
1963	425		520			
1964	362			555		
1965	448	548	600	650	672	
1966	453	718	907	901	947	
1967	551	886	992	1051	1070	1135
1968			1075	1183	1226	1217
1969				1287	1400	
1970					1398	1394
	570	858	1028	1237	1393	1407
	Number of Missiles Actually Built by Russia					

Source: Lawrence Freedman, *U.S. Intelligence and the Soviet Strategic Threat* (London: Macmillan, 1977), 106.

vantages." Apparently his colleagues in the Kremlin also were upset about strong statements Carter had issued on human rights in the Soviet Union and his public style of diplomacy. Talks resumed subsequently in private and on terms more acceptable to Moscow. What finally emerged after two years resembled the Vladivostok guidelines much more closely than had the initial Carter proposal. Important modifications to the Vladivostok agreement included the reduction of the overall ceiling on missiles and bombers from 2,400 to 2,250 launchers, and the addition of limits on outfitting missiles with multiple warheads.

A protocol to the treaty, which was to expire at the end of 1981, banned the deployment or flight testing of a land-based mobile missile. The protocol also barred the deployment, but not the testing, of land- and sea-launched cruise missiles with ranges over 360 miles. A statement of principles, finally, committed each party to begin a new round of strategic arms talks — SALT III — once SALT II was ratified. SALT III was to lead to "significant reductions" in each country's strategic arsenal.

SALT II Limits

MAXIMUM LEVELS PERMITTED

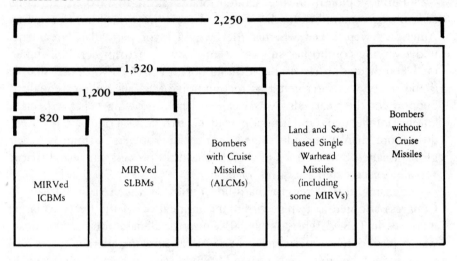

Opponents of the SALT II treaty argued that it would impede the development and deployment of new U.S. weapons, which, they said, were needed to counter the Soviet threat. Opponents especially were concerned about the alleged vulnerability of U.S. land-based missiles and Soviet efforts in anti-ballistic missile defense and civil defense, areas that suggested to them that the Soviet Union was gaining the capability to fight and win a nuclear war. Even if the Soviet Union never would dare launch a pre-emptive strike against U.S. missiles, the mere perception of U.S. vulnerability might embolden Soviet leaders to challenge the United States in regional disputes, the opponents of SALT argued. Most of all, opponents feared that the arms control process was lulling the United States into a false complacency. Eugene V. Rostow, chairman of the Committee on the Present Danger, wrote in 1979 that the arguments in favor of SALT II were the same as those proffered for naval limitation and other arms control agreements in the 1920s. Those agreements, Rostow said, "helped to bring on World War II by reinforcing the blind and willful optimism of the West, thus inhibiting the possibility of military preparedness and diplomatic actions through which Britain and France could easily have deterred the war." [5]

Public opinion in the United States strongly favored an arms control agreement with the Soviet Union in principle. But when Americans were asked whether they would favor particular agreements under varying conditions, support seemed somewhat flimsier. Polls taken in 1979 indicated, for example, that support for SALT II dropped sharply if the treaty was interpreted as recognition of Soviet military equality.[6] Support for the treaty also was weakened by a growing belief, especially among intellectuals and opinion leaders, that arms control talks were little more than a charade staged by the superpowers to improve their global images. Some critics of the arms control process suggested that it actually exacerbated the arms race.[7]

Sensing their strength, hard-line critics of the SALT II treaty in Congress adopted a strategy of linking ratification of the treaty to larger increases in U.S. defense outlays. Because the Senate vote on the treaty was expected to be extremely close, it was anticipated that just a few votes would be able to command tremendous influence. On Aug. 2, 1979, two months after Carter and Brezhnev signed the treaty, Sens. Sam Nunn, D-Ga., John Tower, R-Texas, and Jackson wrote to Carter demanding annual increases in the defense budget of at least 5 percent as the price of their support for the treaty. The preceding March three other senators, George S. McGovern, D-S.D., Mark O. Hatfield, R-Ore., and William Proxmire, D-Wis., had written to Carter warning that they would vote against a treaty that did not "fundamentally curb the arms race."

For three months Carter tried to hold Nunn, Tower, and Jackson at bay. But in December 1979, following the seizure of U.S. hostages in Iran, Carter gave in and announced he would request 5 percent annual increases in defense spending. The following month, after the Soviet invasion of Afghanistan, he withdrew the SALT treaty from consideration. At the time of the Soviet intervention the Carter administration was quietly making preparations for a naval blockade of Iran, but these plans were dropped after Soviet troops moved into Afghanistan, in recognition of the Soviet Union's greater capacity to mount military operations in the area.[8] In 1977-1978 the Soviet Union had airlifted military supplies to a revolutionary government in Ethiopia, highlighting its ability to move equipment and personnel in the Middle East area.

In his State of the Union address in January 1980 Carter tried to address public concern about Soviet activities and capabilities in the Middle East by proclaiming a new "doctrine." The Carter doctrine said that the United States would use whatever armed force was necessary to defend its interests in the Persian Gulf. But by this time, even among people who wholeheartedly supported tough military policies, there was grave skepticism about whether the United States was in a position to enforce Carter's commitment. To many people it seemed a grave mistake to emphasize the importance of U.S. interests in an area where the country was so poorly prepared to defend them with anything short of nuclear weapons.

By January 1980 public opinion had shifted sharply in favor of higher defense spending, and Carter was in serious political trouble. Sensing an opportunity to go for broke, the advocates of a much faster military buildup signed onto the Reagan campaign in growing numbers. Reagan ran on a platform endorsing a "national strategy of peace through strength" to be based on achievement of "overall military and technological superiority over the Soviet Union." In his acceptance speech at the Republican convention in the summer of 1980 Reagan rejected what he called the Democratic view that "the United States has had its day in the sun; that our nation has passed its zenith."

While Reagan opposed reintroduction of the draft and expressed doubts about Carter's basing plan for the MX missile, he favored just about every other weapon and program sought by the military, and he promised to boost defense spending by 7 percent annually if elected president. Carter sought to portray Reagan as a warmonger, but to no avail. In the November 1980 landslide election a number of leading congressional doves went down to defeat with Carter, including Senator McGovern and Sen. John C. Culver, D-Iowa.

Reagan's 7 Percent Solution

Consistent with Reagan's campaign pledge to recover military superiority over the Soviet Union, his administration pushed ahead with and accelerated virtually all major weapons systems that were being built or planned when he took office. The premises of the administration's policy were that the United States could afford to outspend the Soviet Union because the American economy was much larger, and that it would beat Russia in an all-out military competition because the U.S.

Public Opinion on Defense Spending

Between the end of the Vietnam War and the Soviet invasion of Afghanistan in December 1979, U.S. public opinion changed markedly on the question of whether the country was spending too little, too much, or about the right amount on defense. The following figures are from Gallup Polls taken in March 1976 and January 1980.

	Too little	*Too much*	*About right*	*No opinion*
1980 Nationwide Sample	49	14	24	13
East	47	16	25	14
Middle West	52	12	26	10
South	54	13	21	12
West	44	13	26	17
1976 Nationwide Sample	22	36	33	5
East	18	38	33	11
Middle West	16	42	34	8
South	28	31	30	11
West	27	35	31	7

lead was so large in the advanced technologies relevant to sophisticated weapons systems.

The most important strategic weapons systems under development when Reagan took office were the Trident submarine, a giant missile-carrying vessel slated to replace the nation's fleet of Poseidons; the B-1 bomber, an extremely expensive aircraft designed to replace the country's aging fleet of B-52 bombers; the cruise missile, a pilotless jet aircraft that uses computer maps of enemy territory to find targets with virtual pinpoint accuracy; and the MX missile, a large intercontinental rocket designed to replace or complement the Minuteman III missile, which currently is deployed in hardened silos in states west of the Mississippi.

Every one of these systems was controversial when Reagan took office: the Trident because of its high cost and large size, which critics said would make it unnecessarily vulnerable to detection by the enemy; the cruise missile because of its low cost and small size, which critics said could make it difficult or impossible to verify in the context of strategic arms agreements; the B-1 bomber because of its very high cost and vulnerability to attack; and the MX because of its astronomically high cost, its potential impact on the environment, its strategic justification, and its suitability for use in a surprise attack on Soviet nuclear missiles.*

During the Carter years work was proceeding apace on the Trident, and cruise missiles were being developed even though their deployment was to be contingent on the outcome of arms control negotiations with the Soviet Union. Carter persuaded Congress to delete funding for the B-1 bomber, mainly on the ground that it probably would be obsolete by the time it entered service. While Carter went along with the Air Force's plans to build and deploy MX missiles on giant trucks in Utah and Nevada at a cost of at least $30 billion, he did so largely in the hope of winning support for the SALT II treaty, and by the time of his re-election defeat the MX had run into intense opposition both locally and nationally.

Reagan continued with the Trident and cruise missile programs, got Congress to restore funding for the B-1, and obtained funds to begin building the MX even though no agreement could be reached within the executive branch on how the missile might be based. Reagan's aim was to modernize every arm of the U.S. strategic "triad" — the bombers, nuclear submarines, and land-based missiles. In addition Reagan requested added funds for civil defense and the development of anti-ballistic missile systems, arguing that the United States had to match Soviet efforts in these areas.

On March 30, 1982, Reagan proposed a seven-year, $4.2 billion plan for civil defense, which allegedly would enable as much as 80 percent of the nation's urban population to survive a nuclear war. On March 23, 1983, he announced a "comprehensive and long-term research and development program" geared to the earliest possible development of effective anti-ballistic missile (ABM) systems. The objective, Reagan said in a speech to the nation, would be to "break out of a future that relies solely on offensive retaliation for our security."

Critics of the Reagan programs have argued that a crash program to develop an ABM system would undermine the system of "mutual assured destruction" and violate the SALT I accords. Critics concede that the So-

---

* Few critics of the MX actually believe that the U.S. government will ever launch a surprise nuclear attack against the Soviet Union, but among arms control specialists it generally is agreed that the mere possibility of doing so would provoke the Soviet Union to take counter-measures. It is important to note, in this connection, that Russian missiles are much more vulnerable to attack than American missiles. A much higher proportion of the Soviet missiles are based on land rather than submarines, and U.S. missiles are more accurate.

viet Union has put a rather substantial effort into ABM research and civil defense planning, but they question the effectiveness of Russian programs and are not inclined to ascribe sinister motives to Soviet leaders.[9] The question of whether Russian leaders think losses could be kept to an acceptable minimum in a nuclear war has had an especially important bearing on the question of whether to build the MX missile. Proponents of the mobile missile have rested their case on the fact that the Soviet Union would have the theoretical capacity by the mid-1980s to wipe out the entire U.S. land-based missile fleet in a surprise attack. During the 1970s, first the United States then the Soviet Union deployed hundreds of new missiles with increasingly accurate multiple warheads.[10] The new U.S. missiles were more accurate than their Soviet counterparts, as U.S. missiles always had been. Perhaps to compensate for their inferior targeting technology, the Soviet Union continued to build much bigger missiles. But as the gap narrowed between U.S. and Soviet targeting capabilities the possibility arose that the Soviet Union would deploy warheads on their huge missiles that have both the yield and the accuracy to destroy U.S. land-based missiles in their silos. The latest Minuteman III missiles, on the other hand, still did not have the right combination of accuracy and yield to destroy Soviet missiles, even with the addition of the new Mark 12A warheads scheduled for deployment during the first half of the 1980s.[11]

If the entire land-based fleet were destroyed in a surprise attack, the United States still could retaliate with its sea- and air-based nuclear weapons. But according to some MX proponents these sea- and air-based weapons would be accurate enough only to hit Soviet cities — not military targets such as the remaining Soviet land-based missiles and their command centers.[12] Because an attack on Russian cities inevitably would be met with an attack on American cities, MX proponents fear that U.S. leaders would be tempted to surrender to Soviet demands rather than to retaliate. To avoid this choice, they conclude, the United States must have a new missile capable of hitting Russian military targets with adequate accuracy and yield, and the missile must be mobile so as to survive a surprise "first strike" attack.

MX critics argue that this nuclear war scenario is implausible on technical, military, and psychological grounds. While the Soviet Union may attain the theoretical capability to wipe out the U.S. land-based missiles, this capability is based on missile tests that are carried out under conditions very different from those in an actual conflict. The United

States tests its missiles on a range that stretches from east to west, and the Soviet Union on a range stretching west to east. But in a war, said Richard Garwin of the Defense Science Board, each side would fire its missiles north over the pole, and a new set of random factors affecting missile accuracy would come into play, including "anomalies in the earth's gravitational field, varying densities of the upper atmosphere or unknown wind velocities." [13]

If the Soviets were to stage a surprise first strike, therefore, they would be gambling on the hope that they had estimated such random factors not merely with fair accuracy but with virtually perfect accuracy. They would be staking their survival on the flawless execution of a complicated maneuver in which the margin of error would be measured in feet and seconds. To destroy 100 percent of the U.S. missiles, at least a thousand Soviet missiles would have to arrive exactly on target — after traveling some 5,000 miles — at exactly the same time.

Even if the Russians were confident of executing this maneuver perfectly — the critics continue — they still would be gambling on U.S. unwillingness to unleash its massive arsenal of sea- and air-based nuclear warheads. This arsenal includes more than 7,000 hydrogen bombs, easily enough to destroy every significant center of population in the Soviet Union many times over. MX critics conclude that Soviet leaders would have to be crazy to take risks of this order.

A Farewell to Arms Control?

When Reagan took office it was plain that he would put a buildup of U.S. strategic (and conventional) forces far ahead of arms control, and in fact there was some question as to whether Reagan would engage in arms negotiations with the Soviet Union. His appointee to head the Arms Control and Disarmament Agency was Eugene Rostow, who had been chairman of the Committee on the Present Danger, the most influential anti-SALT group in the Ford and Carter years. When the administration entered into talks with Russia about medium-range missiles in Europe, Reagan chose as U.S. negotiator Paul H. Nitze, who had been staff director for the Committee on the Present Danger. When strategic arms talks started a year later Reagan picked as negotiator retired general Edward L. Rowny, who had resigned from Carter's SALT II delegation in protest against concessions made to Russia.

To many observers the negotiating positions that Reagan adopted in arms control talks seemed to be designed more to mollify public

The Geneva Talks...

'START' Talks

The "strategic arms reduction talks" involve the missiles and bombers that each superpower targets at the other. Edward L. Rowny heads the U.S. delegation, Viktor M. Karpov the Soviet delegation.

November 1974: Gerald R. Ford and Leonid Brezhnev sign Vladivostok agreement, which limits each side to a total of 2,400 strategic delivery vehicles (missiles and bombers).

June 1979: Jimmy Carter and Brezhnev sign SALT II treaty at Vienna. Ceilings for each side reduced to 2,250 delivery vehicles; protocol temporarily limits deployment of cruise missiles and mobile intercontinental missiles.

January 1980: President Carter gives up on gaining Senate ratification of the SALT II treaty following Soviet invasion of Afghanistan.

June 29, 1982: START talks open in Geneva.

April 1983: A presidential commission on strategic forces headed by Lieut. Gen. Brent Scowcroft, Ret., recommends restructuring U.S. forces to rely on small, single-warhead missiles. Also recommends proceeding with multi-warhead MX missiles, however.

August 1983: Reagan administration reported to be relaxing some key demands in START talks.

'INF' Talks

The "intermediate-range nuclear forces" negotiations involve missiles and perhaps bombers based in or targeted at Western Europe and European Russia. Yuli A. Kvitsinsky heads the Soviet team, Paul H. Nitze the American.

1977: The Soviet Union begins to deploy SS-20 medium-range missiles.

October 1977: West Germany's Helmut Schmidt, alluding to Soviet deployment of Soviet SS-20 missiles, decries the exclusion of such

. . . A Chronology

systems from SALT negotiations and calls for some kind of corrective action.

December 1979: NATO adopts "dual track" decision to deploy 108 Pershing II and 464 cruise missiles in five West European countries beginning in December 1983, unless an agreement is reached with the Soviet Union limiting such missiles.

November 19, 1981: Reagan proposes the so-called "zero option." He says that the United States will cancel the decision to deploy new missiles in Western Europe if the Soviet Union dismantles all SS-20, SS-4, and SS-5 missiles.

November 30, 1981: INF talks open in Geneva.

March 16, 1982: Soviet leader Leonid Brezhnev announces freeze on further deployment of SS-20 missiles in European Russia.

July 1982: "Walk in the woods" agreement: Nitze and Kvitsinsky informally explore the idea of an agreement in which NATO would forgo deployment of Pershing II missiles; SS-20 and cruise missiles would be subject to an equal limit of 75. Idea vetoed by both governments.

September 1983: Positions harden following the downing of a commercial South Korean airliner by Soviet pilots. Russian leaders say they will not adopt a more conciliatory position in arms control talks to placate world opinion, and they back off from conciliatory offers made late in the summer.

December 1983: The Soviet Union suspends both INF and START talks after deployment of U.S. Pershing II and cruise missiles begins.

opinion than to enhance prospects for serious bargaining. By early 1983 the question of whether the U.S. government was negotiating in good faith in Geneva had become a significant political issue in a number of European countries. Such concern became especially acute in January when Reagan fired Rostow — apparently for not adhering closely enough to the administration's tough bargaining positions — and replaced him with Kenneth Adelman, a 36-year-old member of the U.S. delegation to the United Nations who had little background in the technicalities of arms control. Later in 1983 it was reported that Reagan had rejected a proposed compromise worked out by Nitze in the talks on medium-range missiles, even after Nitze told him that the basic U.S. negotiating position was unrealistic. "Well, Paul," Reagan is reported to have said, "you just tell the Soviets that you're working for one tough son of a bitch." [14]

In December 1983, when the United States began to deploy medium-range Pershing II and cruise missiles in Europe over the Soviet Union's strenuous objections, the Russians broke off all arms control talks. They announced their intention of deploying medium-range missiles in East Germany and Czechoslovakia, and in January 1984 U.S. intelligence sources confirmed that they already were stationing a new medium-range missile — the SS-22 — in East Germany.

An 'Arms Race?'

Since 1945 neither of the superpowers has felt free to lag far behind the other in the development and deployment of nuclear weapons. Yet at the same time both have stopped short of mobilizing their resources singlemindedly for war. None of the U.S. or Soviet military buildups since 1945 has come close to resembling Hitler's all-out mobilization of the German economy before World War II.

During the late 1940s and mid-1950s both U.S. and Soviet leaders tried hard to economize on overall military spending, though both countries also worked hard to develop long-range missiles and hydrogen bombs. In the 1960s, first the United States (under Kennedy) and then the Soviet Union (under Nikita Khrushchev's successors) opted for larger and more comprehensive increases in defense spending. According to the latest CIA estimates, which the Joint Economic Committee of the U.S. Congress released in November 1983, the Soviet Union boosted its military outlays by an average of 4 to 5 percent a year from 1966 to 1976. After that, the CIA concluded (contrary to its estimates in the late 1970s), the Soviet Union trimmed its annual increases in defense spending to around 2 percent a year.

At the beginning of 1984 the CIA estimated that the Soviet Union was allocating 13 to 14 percent of its gross national product to military spending, about twice the proportion of U.S. annual production that was going for defense. The U.S. economy is about twice as large as the Soviet Union's, however, and so the United States does not have to spend so high a proportion of its national product to keep up with the Soviet military efforts.

The military rivalry between the superpowers often is described as an action-reaction cycle in which each side has felt compelled to match instantly any of the other side's moves. But to a great extent each side has felt relaxed enough to determine its arms policies from year to year somewhat independently of the other. While an action-reaction pattern is apparent, the key policy changes have tended to occur once a decade or so, not every year. In the early 1970s, for example, the United States answered the Soviet missile buildup of the late 1960s by equipping its ICBMs with multiple warheads. When the Soviet Union responded by putting multiple warheads on its missile fleet in the late 1970s, U.S. strategic planners began to worry about the growing vulnerability of American land-based missiles, and pressure grew for deployment of a mobile land-based missile like the MX. In a sense the buildup of U.S. forces that began around the time Carter took office in 1977 has been a reaction to the Soviet buildup that took place between 1966 and 1976 when the United States was trying to cut non-Vietnam-related defense spending.

In recent years the quantity of nuclear delivery systems maintained by each superpower has remained relatively constant, but both countries have made energetic efforts to improve the quality and effectiveness of their weapons. Since the late 1970s technological advances have threatened to undermine the foundations of arms control and bring on a heightened peril of nuclear war. Generally, nuclear weapons systems have tended to become more compact and easily concealed, more accurate, and more precise in their effects. As a result public officials, politicians, and independent defense experts have begun to talk more and more about the possibility of actually using such weapons to promote military objectives.

In the controversies that developed over nuclear weapons in the late 1970s and early 1980s arguments almost always ended up turning on the question of "nuclear war fighting" or "counterforce" — the use of nuclear weapons against military targets to achieve limited objectives (short of totally annihilating the adversary). According to one school of thought, the more prepared a country is to fight and win a nuclear war, the less likely it is that the adversary will take measures that result in such a war. According to the opposing school, the more leaders talk about fighting nuclear wars and the harder they try to acquire weapons suitable for use in such wars, the more likely it is that somebody eventually will start one.

In their relations since 1945 the leaders of the superpowers plainly have believed that nuclear weapons should be used only as an absolute last resort. But there have been extremely serious confrontations in which the use of nuclear weapons seemed to be a live possibility. And leaders on both sides continue to insist that the nuclear balance and perceptions of the nuclear balance figure in the way regional disputes are resolved.

Notes

1. *Congressional Quarterly Weekly Report*, January 6, 1948, 1.
2. See Robert A. Divine, *Blowing in the Wind — The Nuclear Test Ban Debate, 1954-1960* (New York: Oxford University Press, 1978), 94-96.
3. *Congressional Quarterly Almanac, 1960* (Congressional Quarterly, 1961), 776.
4. See Cecil V. Crabb, Jr., and Pat M. Holt, *Invitation to Struggle: Congress, the President and Foreign Policy* (Washington D.C.: CQ Press, 1980), 126-127, 145-147.
5. Eugene V. Rostow, "The Case Against SALT II," *Commentary* (February 1979): 30.
6. Generally, two-thirds to four-fifths of the public supported the idea of a new arms control treaty with the Soviet Union. But in one poll, for example, support for a treaty dropped to 50 percent when people were asked whether the United States should agree to a treaty that left it merely equal rather than superior to the Soviet Union. See *The Washington Post*, May 25, 1979.
7. See Alva Myrdal, *The Game of Disarmament* (New York: Pantheon Books, 1976); Sverre Lodgaard, "Functions of SALT," *Journal of Peace Research*, v. XIV, no. 1 (1977): 1-22 and Jane Sharp, "Swords and Plowshares," *Arms Control Today* (June 1977).
8. See Zbigniew Brzezinski, "The Failed Mission," *The New York Times Magazine*, April 18, 1982, 28.

9. According to a 1978 CIA study, the Soviet civil defense effort involves roughly 100,000 full-time employees. Unlike the United States, which has constructed no blast-resistant shelters, the Soviet Union probably has built around 15,000, enough to house "roughly 10 to 20 percent of the total population in cities of more than 100,000 people," the CIA study said. "Nearly every Soviet citizen receives civil defense instruction," the CIA found, but it uncovered "no evidence that evacuation exercises in large cities involving actual movement of people have been practiced." A study prepared by the U.S. Arms Control and Disarmament Agency in 1978 found that Russian civil defense planning stood little or no chance of preventing nuclear war from being catastrophic. Soviet leaders frequently have said that tens or hundreds of millions of people would die in a nuclear war, and estimates of probable casualties by physicians have been widely reported in the Soviet press. See David Holloway, *The Soviet Union and the Arms Race* (New Haven: Yale University Press), 52-53, 62-63.

10. Between 1970 and 1974 the United States modernized its land-based missile fleet with 540 new Minuteman III missiles and more than 1,000 additional warheads. During this period the Soviet Union made relatively insignificant changes in its land-based missile arsenal. But between 1975 and 1979 the Soviet Union deployed 620 SS-17, SS-18, and SS-19 missiles and more than 2,000 additional warheads. These are the weapons that most seriously concern U.S. defense planners. See John M. Collins, *U.S.-Soviet Military Balance* (New York: McGraw-Hill, 1980), 443. Other authorities claim that the Minuteman III missiles armed with MK-12A warheads are in fact capable of attacking hardened Soviet silos. See Thomas B. Cochran, William N. Arkin, and Milton M. Hoenig, *Nuclear Weapons Data Book* (New York: Harper & Row, 1984).

11. Ibid., 133.

12. According to Harold A. Feiveson and Frank von Hippel, the bombs and cruise missiles carried by U.S. planes are accurate enough to destroy hardened silos, and submarine-launched missiles are accurate enough to be targeted at military installations such as submarine bases and airfields. See their letter to *Physics Today* (May 1983): 13.

13. Quoted in Andrew Cockburn and Alexander Cockburn, "The Myth of Missile Accuracy," *The New York Review of Books*, Nov., 20, 1983, 43

14. See Strobe Talbott, "Behind Closed Doors," *Time*, Dec. 5, 1983, 18-37; also, John Newhouse, "Arms and Orthodoxy," *The New Yorker*, June 7, 1982, 44-103.

On October 24, 1981, a crowd of 100,000 participated in the Campaign for Nuclear Disarmament's march through London's West End. Similar marches were held in other parts of Europe throughout the early 1980s.

Chapter 11

THE INTERNATIONAL PEACE MOVEMENT
AND THE SUPERPOWERS

In 1980-1981 a grass-roots movement against nuclear weapons emerged first in Europe, then in the United States. American anti-nuclear activists rallied around a campaign for a bilateral freeze on the testing, production, and deployment of new nuclear weapons. European activism centered primarily on the deployment of U.S. medium-range missiles in Germany, England, Italy, the Netherlands, and Belgium, which began at the end of 1983. But the European movement has deeper roots, and its members have raised fundamental questions about the division of Western civilization into two hostile blocs and the reliance of each bloc on nuclear deterrence.

Since the late 1950s European anti-nuclear activism has been intimately connected with what Harold A. Feiveson of Princeton University has called the "dilemma of theater nuclear weapons" (nuclear weapons based in or targeted at Europe).[1] When large numbers of "tactical" or "battlefield" nuclear weapons first were deployed in Europe during the late 1950s, civilian leaders had little notion of how such weapons actually might be used in combat. NATO strategists developed a theory of "graduated" or "controlled" response, which held that first conventional and then tactical nuclear weapons would be used to repulse a Soviet attack before resorting to massive retaliation. To some Europeans, this so-called flexible response doctrine opened the possibility that the United States would stop short of ever using nuclear weapons directly against the Soviet Union, even if the Red Army were winning a conventional or tactical nuclear war in Western Europe. To others, flexible response held up the terrifying prospect that the superpowers might fight a "limited nuclear war" on European turf while keeping Russian and American territory out of the action. To still others it

Deployment of Nuclear Weapons . . .

December 1954 NATO adopts policy of countering Soviet attack with tactical nuclear weapons.

1955 "Carte Blanche" tactical nuclear maneuvers in West Germany; deployment of first Soviet medium-range missile, the SS-3, begins.

1957 United States agrees to supply England with 60 Thor medium-range missiles; decides to deploy missiles in other European countries, but in due course only Italy and Turkey agree to accept them.

1959 Deployment of Soviet SS-4 missile begins.

1961 Deployment of SS-5 begins.

June-July 1962 Defense Secretary Robert McNamara outlines "controlled response" doctrine, indicating that the United States would rely first on conventional forces, then tactical nuclear weapons, and only finally on strategic nuclear weapons in a confrontation with Russia; Jupiter missiles in Turkey become operational.

December 1962 United States agrees to supply England with Polaris (submarine-launched) missiles.

January 1963 De Gaulle rejects U.S. offer of Polaris missiles.

April 1963 Jupiter missiles withdrawn from Turkey.

seemed far-fetched to imagine that the superpowers would be able to control the escalation of a war once nuclear weapons of any kind had been used.

Soviet leaders never have accepted the notion that a European war could be limited, and when NATO deployment of tactical nuclear weapons began the Soviet response was to concentrate on deployment of medium-range missiles targeted at Europe. This helps explain why the Soviet Union was so far behind in the race to deploy long-range missiles in October 1962 when the Cuban confrontation took place.[2] If a war broke out in Europe, Soviet doctrine as of February 1984 called for immediate nuclear strikes on military targets in the West to pave the way for a tank assault.[3]

...in the European Theater

March 1966	De Gaulle announces French withdrawal from NATO command.
1967	NATO adopts "flexible response" doctrine.
1975	France begins deployment of Pluton tactical missile.
1977-1978	Controversy over proposed deployment of U.S. neutron bombs in West Europe.
1977	Soviet Union begins deployment of SS-20 mobile missile.
December 1979	NATO decides to deploy Pershing II and cruise missiles beginning in 1983, but to seek negotiations with Russia in the meantime; stockpile of tactical nuclear weapons in West Europe to be reduced to 6,000 from 7,000.
July 1980	United States agrees to provide England with Trident (submarine-launched) missiles.
October 1983	NATO decides to reduce stockpile of tactical nuclear weapons to 4,600 from 6,000.
December 1983	Deployment of Pershing II and cruise missiles begins; Soviet Union suspends arms control negotiations.
June 1983	Defense Secretary Caspar Weinberger announces that the Soviet Union had been deploying tactical nuclear warheads in East Europe "since about 1979."

In recent years theater nuclear weapons have come under criticism for a large number of pragmatic reasons. They have been called a magnet for Soviet nuclear strikes, an obstacle to rational defense planning and — plain and simply — unneeded. High-ranking NATO generals have argued, for example, that nuclear artillery shells, missiles, and bombs tie up troops as well as dual-capable howitzers and aircraft.[4] The Warsaw Pact — the East European defense organization — has deployed many more tanks than NATO, but critics of theater nuclear weapons argue that armored vehicles have become exceedingly vulnerable to precision-guided weapons. In preparation for the 1973 Middle East War, the Arab states are said to have lined up about as many armored vehicles as Hitler used in his attack on the Soviet Union in 1941; yet all their tanks were stopped by the Israelis before they had advanced 10 miles.[5]

Ultimately, as Feiveson has argued, theater nuclear weapons embody and sharpen a moral paradox that is at the heart of the deterrence doctrine. The threat to employ such weapons may be a rational means of preventing war, especially when faced with adversaries who may have superior conventional forces at their disposal. But actual use of such weapons might be irrational and immoral in that it immediately would cause millions of deaths on and near the battlefield and most likely provoke an all-out nuclear holocaust.

Thus, Feiveson observed, the deterrence doctrine

> leads to a crisis of democracy. For logic suggests that citizens should vote for those leaders who appear to be without scruples — for the less humane. It may be both consistent in some sense and moral for a political leader to threaten nuclear annihilation, and yet immoral for him actually to implement the threat; and perhaps, in his heart of hearts, he intends not to. But this condition implies an intolerable juggling act on the part of the citizenry — to elect leaders who will appear to the adversary to be without scruples, but who in a crisis will turn out to have them after all.[6]

European Anti-Missile Movement

Anti-nuclear activism first appeared in Europe during the late 1950s. In 1958 a newly formed organization in England called the Campaign for Nuclear Disarmament (CND) organized mass demonstrations to "ban the bomb." In the process, the CND devised a symbol that has since become the badge of peace activists all over the world. In Germany the Social Democratic Party (SPD) rallied demonstrators to back its anti-militarist, anti-nuclear, and neutralist foreign policies.

The early anti-nuclear protests proved short-lived, in part, perhaps, because the dissenting leaders frequently took positions that seemed excessive or simple-minded. British activists often said that it would be better to be "red than dead," which seemed to imply that the price of giving up nuclear weapons very well might be Soviet military occupation and forced conversion to communism. SPD leaders argued that a neutral West Germany eventually could be re-united with East Germany, even though to many people it seemed highly improbable that the Soviet Union — not to speak of West European countries — would permit German unification under any circumstances.

Activism against nuclear weapons first appeared in the 1950s when Americans demonstrated for a ban on tests of atomic weapons, Germans protested military exercises, and the British turned out for protests in favor of unilateral nuclear disarmament. The British demonstrations were organized by the Campaign for Nuclear Disarmament (CND), which adopted a symbol derived from the semaphore sign for ND — nuclear disarmament. The symbol consists of a clock with the arms at four, eight, and 12, and it later was adopted by peace activists all over the world, regardless of whether their causes had anything to do with nuclear weapons.

When student demonstrations swept European cities in 1968, criticism of nuclear policies played little or no part. Already apparent, though, was a desire among young Europeans to be more independent of both superpowers and even-handed in criticism of both countries. In the first uprisings, which started in March 1968 at Warsaw University, Polish students — soon joined by workers — expressed solidarity with the Vietnamese communists, Czech reformers, and Soviet dissidents. In Germany, student radicals increasingly styled themselves as Maoists, which may have served as a way of striking a radically critical pose toward the United States without being associated with Soviet communism. In April 1968, following an unsuccessful attempt on the life of

Rudi Dutschke, a leader of Germany's Student Federation of German Socialists (SDS), student riots broke out in Berlin and spread to Hamburg, Frankfurt, Munich, Hanover, Essen, Cologne, Bremen, Stuttgart, and Heidelberg. The students were especially critical of the West German government for not speaking out against U.S. policies in Vietnam, and among those arrested was the son of the incumbent socialist chancellor, Willy Brandt. The European protests came to a climax in May 1968 when French students erected street barricades and fought pitched battles with Paris police after President Charles de Gaulle's government barred student leader Daniel Cohn-Bendit from re-entering France from West Germany. Cohn-Bendit was the author of a book on "new left" ideology in which he presented a "left-wing alternative" to the Soviet Union's bureaucratic "state capitalism." [7]

The European protests came to an end in August 1968 when Soviet tanks rolled into the streets of Prague, where the government had been trying, contrary to the wishes of the Soviet leadership, to liberalize their political and economic systems. In France the government instituted university reforms that went a long way toward accommodating student grievances. In Germany, where the older generation's authority was weakest because adults were associated with Hitler, the concentration camps, and a losing war, some student radicals became terrorists. Others heeded Dutschke's call to make a "long march through the institutions," in other words, to pack up their Mao jackets, don shirts and ties, get jobs, and try to exert their influence through established institutions. In Poland a deceptive calm returned, but in 1976 students and workers once again joined in a rebellion against the communist authorities. And in the summer of 1980 the same rebel alliance temporarily got the best of the government and proceeded — for one year — to force far-reaching social, political, and economic reforms.

By this time a full-fledged anti-nuclear weapons movement had appeared in West Europe, arousing mixed feelings on both sides of what used to be called the "Iron Curtain." The first signs of such a movement appeared in the early and mid-1970s when large-scale demonstrations took place at nuclear reactor construction sites in Germany and France. Many of the German demonstrators were organized by so-called "citizens' initiative" or citizens action groups, which had sprouted up by the hundreds and thousands since the late 1960s. By the late 1970s some 1,000 groups of this kind with more than 300,000 members were joined together in the Federal Association of Citizen Action Groups for the

Environment (BBU), which had been founded in 1972. Meanwhile, veterans of the 1968 demonstrations were becoming increasingly influential in the left wing of the SPD.

In 1977-1978 when the Carter administration was making plans to deploy neutron bombs in West Europe, demonstrators began to turn out for the first time since the late 1950s and early 1960s to protest strictly against nuclear weapons and nuclear weapons policies. The Dutch communist party, which is independent of Moscow's control but not of its influence, launched the first campaign against the neutron bomb. Because the neutron bomb is designed to kill combatants but spare civilians near the battlefield, it was easy for propagandists to describe the "enhanced radiation warhead" as a weapon designed to kill people but spare property. Egon Bahr, the top disarmament expert in Germany's SPD, called the neutron bomb a "symbol of mental perversion." [8]

By the end of 1977, Holland's Inter-Church Peace Council (IKV) had joined the campaign against the neutron bomb and replaced the communist party as the most influential European organization behind the protests. In Germany the bomb had become a highly divisive issue within the SPD, and Chancellor Helmut Schmidt told President Jimmy Carter that he would accept the warhead only if at least one other continental country did so as well. Complicated negotiations ensued among the NATO powers in which the Europeans took the position that deployment of the neutron bomb would be accepted only if 1) the decision to produce the weapon was purely the U.S. government's and 2) attempts were made to win concessions from the Soviet Union in exchange for curtailed production. Carter, finding this position insufficiently supportive, cancelled production but not development of the weapon in April 1978.

In December of the following year an even more divisive nuclear weapons issue began to plague the alliance when the NATO countries decided — two weeks before the Soviet Union's invasion of Afghanistan — to deploy nearly 500 Pershing II and cruise missiles beginning in December 1983. The NATO decision was a response to Soviet deployment of the SS-20, which began in 1977, and it was taken only on the condition that the United States would seek negotiations with Russia to limit missiles located in or targeted at the European theater. In talks with the U.S. government, moreover, Chancellor Schmidt said that his government would accept stationing of the missiles on German soil only if two other European countries did so as well. The missiles posed especially sensitive questions for Germany because for the first time since World War II there would be nuclear missiles on German soil capable of striking at the Soviet Union's heartland.

After Ronald Reagan took office in 1981, anti-missile activism rapidly gathered force in Europe, largely because of a perception that the new president was a "trigger happy cowboy," who might easily involve Europe in a nuclear war. In 1980 and 1981 membership in England's CND grew tenfold (to 30,000), and the Labor Party adopted positions in favor of unilateral disarmament and against missile deployment. In Germany, the SPD's anti-nuclear left wing steadily gained strength, and a new anti-nuclear party — "the Greens" — began to pick up a significant numbers of votes in state elections. Petra Kelly, a member of the BBU environmental coalition, played a leading role in founding the Greens, and one of the party's first victories was in Lower Saxony where mass demonstrations had taken place against government plans to build a large nuclear fuel reprocessing complex. Other early victories took place in Hamburg and Bremen, where independent leftists such as Dutschke campaigned for the Greens.[9] In fall 1981 a wave of massive anti-nuclear demonstrations swept West Europe, and while protestors often took care to emphasize their distaste for all nuclear weapons, they generally left little doubt that they found U.S. arms policies even more offensive than the Soviet Union's. On October 10 about a quarter of a million people turned out in Bonn; on October 24, 250,000 joined in a similar protest in London; the following day, 300,000 marched in Brussels, 250,000 in Rome, 50,000 in Paris, and 10,000 in Oslo. An October 31 demonstration drew 100,000 in Milan, and a protest on November 21 in Amsterdam attracted 300,000. On November 15 a half million people demonstrated in Spain against the country's proposed membership in NATO, and a month earlier the Greeks had elected a socialist premier who had threatened repeatedly to pull Greece out of NATO and to remove all nuclear weapons from its soil.

After Poland's military imposed martial law on Dec. 13, 1981 — obviously with the foreknowledge and encouragement of the Soviet government — it was widely expected that West European anti-nuclear protests would die out, much as the student rebellions had ended in 1968 after the invasion of Czechoslovakia. But in June 1982 yet another large demonstration took place in Bonn. In March of the following year the Greens got enough votes to take seats in Germany's parliament. While the election produced a sharp defeat for the SPD, just as England's

Labor Party had suffered a setback when the British went to the polls in June 1983, economic issues were of decisive importance, and neither party was shocked into switching positions on the missile issue. In November 1983, days before the first U.S. missiles were scheduled to arrive in West Germany, an SPD party congress adopted a resolution calling for a delay in deployment to allow time for further negotiations.

Interpretations of the European Movement

In efforts to account for the emergence of the European anti-nuclear movement, commentators often stress statements by Reagan and top administration officials in which they seemed to indicate rather too casually that they considered limited nuclear war plausible.[10] Commentators often suggest that difficulties with Europe could be avoided if U.S. officials were more careful with their language. While this may be true to an extent, it is important to remember that European protests are directed at the substance of U.S. policy, not the mere words in which it is cloaked. The U.S. defense secretary's guidance plan for 1985-1989 instructed the armed services to "integrate plans for employing medium- and long-range nuclear weapons to enable them to fight a war more effectively," as *The New York Times* reported on March 22, 1983. The Fiscal Year 1985-89 Defense Guidance plan reaffirmed and strengthened a policy that every president has endorsed since Eisenhower.

Reports on the European peace movement by the U.S. press often stress that the university-aged people who turned out for demonstrations were born long after the Berlin airlift and the Marshall Plan; their perceptions of the United States, it is said, are more strongly shaped by the Vietnam War and Watergate. It is not the case, however, that young Europeans are blindly anti-American, uncritical of the Soviet Union, or poorly educated in general. Steven Szabo, the editor of a book on what often is called Europe's "successor generation," has stressed in his work that university enrollment roughly tripled in Europe between the early 1960s and early 1980s. Szabo believes that the result has been a greater feeling of political efficacy and self-confidence, less deferential attitudes toward authority, and a predisposition to think that young people should have a role in shaping decisions.[11]

In Germany, where the gap between the pre-war and post-war generations is sharpest, the conservative press often portrays young people as "cry babies" who grew up in Willy Brandt's and Helmut Schmidt's welfare state where they got everything free without being

U.S. Limited Nuclear War Doctrine

Even in the 1950s, when official U.S. policy called for "massive retaliation" in the event of a Soviet attack on West Europe, the Strategic Air Command's targeting plans called for nuclear strikes against purely military targets. In July 1962 Defense Secretary Robert McNamara outlined a new doctrine of "controlled response" in a speech at Ann Arbor, Mich., and five years later NATO adopted flexible response as its official policy. In 1974 Defense Secretary James Schlesinger reaffirmed that a major U.S. objective was to be able to fight a "limited nuclear war" effectively. In testimony to Congress he claimed that civilian casualties in such a war could be held to the hundreds of thousands, rather than tens or hundreds of millions. This claim, which specialists disputed, provoked a storm of controversy.

President Jimmy Carter took a number of measures to enhance the country's nuclear war fighting capabilities. In September 1978 he signed Presidential Directive 41, which authorized the government to plan for "crisis relocation" of the country's urban populations in the event a nuclear war broke out. Presidential Directive 53 of 1980 directed officials to secure the nation's telecommunications network against nuclear attack to "support flexible execution of retaliatory strikes and after an enemy nuclear attack." Presidential Directive 59 of 1980 gave a higher priority to Soviet military targets and placed especially great emphasis on destroying the Soviet Union's political leadership (an important departure from McNamara's targeting policies, which exempted the Soviet Union's leaders from attack so that they would be available to negotiate an end to a war). Presidential Directive 58 of 1980, issued in connection with Presidential Directive 59, outlined new measures to provide for political continuity in the United States following a nuclear attack.

Ronald Reagan and his advisers, early in his term, were outspoken about adopting policies designed to ensure that the United States could "prevail" in a nuclear war. But such policies did not represent a sharp departure from previous U.S. doctrine. In May 1982 *The New York Times* published parts of a five-year defense plan prepared by the Reagan administration that allegedly confirmed U.S. preparations for a protracted nuclear war.

asked to bear responsibility or make contributions in return.[12] Even Schmidt himself has said that while young Germans "live in great material well-being," they "have contempt for this, but they also claim it is very much due them." The young may have "reached their majority," Schmidt said, but their "mature years" come much later.[13]

Many of the anti-nuclear protestors who gave Schmidt trouble were born in the early 1960s when Europe experienced a delayed baby boom. But Europeans of all ages have turned out for demonstrations, and the leadership of the European anti-nuclear movement consists overwhelmingly of people from the generation of 1968 — men and women who are now in their late 30s, often parents of teen-age children, and altogether well-established members of society. This is true of the left-wing Social Democrats in Germany, the Greens, England's CND, and quasi-religious organizations such as the IKV and "Aktion Sühnezeichen," the German peace organization that commands the widest respect. Aktion Sühnezeichen, which translates literally as "atonement action," was founded after World War II by Lutherans who regarded themselves as spiritual heirs to the "Confessing Church" — a small group of Protestants, led by Martin Niemoeller and Dietrich Bonhoeffer, who refused to keep silent in the face of Hitler's crimes.

The failure of Germany's Lutherans to speak out more strongly against Hitler left many with guilty consciences, and many of those who currently participate in the peace movement regard their activities as a way of doing penance. As early as the late 1950s, Germany's Lutheran church adopted the position that while Christians could "still" justify reliance on nuclear deterrence, it was incumbent on them to press hard for total nuclear disarmament. Today Germany's Protestants have come to work closely in the peace movement with Dutch Reformed congregations, which have a long tradition of activism in civic affairs.

Germany's Catholic Church, like France's, has taken a guarded position against nuclear deterrence despite Pope John Paul II's strong statements against the arms race and militarism. Pope John XXIII in his famous encyclical *Pacem in Terris* (Peace on Earth) wrote that in "this age which boasts of its atomic power, it no longer makes sense to maintain that war is a fit instrument with which to repair the violation of justice."

The European protest movement in its early days often was described as an essentially North European, Protestant phenomenon. But in the demonstrations that swept Europe in fall 1981, Easter 1983, and the "hot autumn" that preceded missile deployment in December 1983,

Peace Research in Europe . . .

To many Americans, Europe's peace movement seemed to come out of thin air. This is partly because there is little public awareness in the United States of the work being done at European peace research institutes. The best known of these institutes is the Stockholm International Peace Research Institute (SIPRI), which has become the world's leading independent authority on global military trends.

Prestigious institutes also are located in Oslo (Norway), Tampere (Finland), Groningen (Holland), Hamburg, Frankfurt, and a number of other European cities. The peace research community is truly European in scope and includes among its most influential members Dieter Senghaas (West Germany); Mary Kaldor (England), daughter of Nicholas Kaldor, the Hungarian Nobel laureate in economics; Alva Myrdal (Sweden), wife of Gunnar Myrdal, the Swedish Nobel laureate in economics; Raimo Väyrynen (Finland), founder of the Tampere Peace Research Institute; Sverre Lodgaard (Norway), director of the Oslo Peace Research Institute (PRIO); and the grand man of the movement, Johan Galtung (Norway), founder of PRIO.

Although Galtung is unknown in this country outside of a small circle of academic specialists, some of his admirers regard him as one of the leading social scientists of the post-war period. Kenneth Boulding, the noted American economist and peace researcher, has compared Galtung to Pablo Picasso. In Boulding's estimation, Galtung is a man who writes "in

turnouts frequently were as large in the South European, Catholic countries as they were in the North. Demonstrations were smallest, ironically, in France and the Scandinavian countries — nations that have contributed most, arguably, to the intellectual foundations of the peace movement. In the domain of strategic theory, de Gaulle was a sharp critic of flexible and limited nuclear war doctrines, while the Scandinavian peace research institutes have been instrumental in mustering facts and arguments to counter the arms race.

It is a striking and suggestive fact that criticism of the United States is most muted — and concern about the Soviet buildup strongest — in countries that were most hostile toward U.S. policies in the period of

. . . An Ongoing Movement

English [not his native language] . . . with a fluency, style and clarity which would be the envy of those who learned English at their mother's knee" and whose "output is so large and varied that it is hard to believe that it comes from only one person." Most of Galtung's ideas are controversial, including the one for which he is best known, the theory of "structural violence" — an attempt to quantify the effects of social injustice so that they can be compared, analytically, to the ravages of war. Galtung himself has changed his mind many times on fundamental matters, and his less fervent admirers say that while he is always interesting, he's always wrong.

According to Väyrynen, who studied at PRIO, the early peace researchers "emphasized conflict theory, decision making, and public opinion in foreign relations, arms control and disarmament as well as nationalism and diplomacy," but by 1970 they were tending to become more oriented to action and protest. Väyrynen has produced estimates indicating that the United States contains about the same number of research institutes as Europe, but he may be including groups that do "peace research" in European terms but are not regarded as peace institutes in the United States. Cornell, Harvard, and Stanford have prestigious arms control programs, and organizations such as the Center for Defense Information in Washington, D.C., produce critical, action-oriented research on U.S. military policy.

American nuclear superiority. It also bears noting, though, that these countries took the greatest pains to remain militarily independent of the United States and now are relatively invulnerable to the vagaries of U.S. electoral politics.

The Freeze Movement

The inspiration for the European uprisings of 1968 was thought by many to have come from the American civil rights and anti-war movements of the early 1960s. In 1979-1980, on the other hand, when the freeze movement began to take root in the United States the spark crossed the Atlantic from the other direction. Randall Forsberg, a scholar and activist who launched the freeze campaign, served her apprenticeship in the field of nuclear weapons as a researcher and writer with the Stockholm International Peace Research Institute. In January 1980, in a meeting at the headquarters of the Fellowship of Reconciliation in Nyack, N.Y., Forsberg persuaded representatives of leading U.S. peace groups to join in a campaign for a bilateral freeze on the development, testing, production, and deployment of new nuclear weapons. The Fellowship of Reconciliation is the American division of an international peace organization headquartered in Holland, which was founded after World War I to promote reconciliation between the Germans and the French. Its German headquarters is in Martin Niemoeller's former home in West Berlin.

When Forsberg formulated the freeze proposal she was deliberately looking for something a little less drastic than the aims pursued by the campaign for European nuclear disarmament — a proposal that would be more appealing to mainstream American opinion. The results exceeded her expectations and the expectations of many who were involved in the decision to launch a freeze campaign. From the end of the Vietnam War to 1979-1980, organizations such as the American Friends Service Committee (AFSC), Clergy and Laity Concerned, and the War Resisters League had been promoting nuclear arms control and disarmament. But their efforts attracted little notice at the time. Terry Provance, the coordinator of the AFSC's campaign against the B-1 bomber and then its disarmament coordinator, supported the freeze campaign from the beginning, but he had no idea how quickly it would catch on — as he remarked in an interview years later.

The movement's breakthrough came in November 1980 when three state senatorial districts in western Massachusetts endorsed the Forsberg freeze proposal. In December of the following year the freeze campaign established a national clearinghouse in St. Louis, Mo., under the direction of Randy Kehler, who had helped mobilize freeze supporters in Massachusetts. By April 1982 town meetings all over New England had endorsed the freeze, as had 33 big city governments and seven state legislatures.

When a freeze resolution was narrowly defeated in the U.S. House of Representatives in mid-1982, groups favoring arms control such as the Council for a Livable World, SANE, and the National Committee for an Effective Congress began to channel funds to defeat freeze opponents and elect candidates favoring the measure. They had some success in the

1982 congressional elections,[14] and pro-freeze referendums passed in eight states and the District of Columbia. Wisconsin voters already had endorsed the freeze in September. On May 4, 1983, the House adopted a slightly modified version of the freeze resolution by a vote of 278-149. The amended resolution said the freeze should be revoked if it was not followed by negotiated arms reductions "within a reasonable, specified period of time." The Senate remained opposed to the freeze by a sizable margin, and more than a year before the 1984 election, arms control groups already were raising money by direct mail to help defeat the senators who did not support their proposal.

In the congressional debates about the freeze resolution, it was at times embarrassingly apparent that many members of Congress — including some outspoken supporters of the move — had inconsistent and conflicting notions of what the move actually would mean. The Forsberg freeze proposal consists of four inter-connected arms control measures: 1) a ban on the production of critical materials (plutonium and highly enriched uranium) for bombs; 2) a freeze on further production of nuclear bombs and warheads; 3) a comprehensive nuclear test ban; and 4) an end to deployment of new weapons systems. Some of these measures, such as the test ban and the prohibition of fissionable materials for weapons, were old staples in the arms control agenda. Others, such as the freeze on new weapons systems, represented a fresh approach to matters that had been or were in the domain of the SALT and START talks.

Of all the aspects of the freeze, the proposed ban on new delivery systems has seemed the most problematic. The intent behind the measure is to prohibit further development and deployment of counter-force weapons, which leaders might be tempted to use in an effort to win a nuclear war.[15] But it might not be feasible to get an agreement that absolutely bars any replacement of existing delivery systems because such an accord would doom the superpowers to total disarmament by obsolescence. If some replacement of weapons were allowed, on the other hand, then difficulties would arise in connection with drawing the line between replacing weapons and building essentially new ones. Difficulties also come up with regard to dual-capable weapons — aircraft, for example, that can carry nuclear or conventional bombs.

Critics of the freeze often charge that it would not be verifiable, but proponents of the measure vigorously dispute this contention. A ban on production of fissionable materials could be verified, they point out, by extending the International Atomic Energy Agency's (IAEA)'s safeguard system to cover the countries with nuclear weapons and by using satellite intelligence to discover circumvention of safeguards. Steps in that direction already have been taken. Over the years, detailed technical means of monitoring compliance with a comprehensive test ban have been worked out in negotiations. No doubt it would be difficult to monitor perfectly the growing numbers of certain new weapons systems, notably the smaller ones such as cruise missiles and howitzer shells. But these would be difficult to monitor in the context of any arms control agreement, and freeze supporters say that a total ban would be easier to verify than the ceilings established in the SALT and START agreements. In a complete ban, it would be enough to detect any evidence of a barred weapon to determine that a violation had taken place.

The idea of a nuclear weapons freeze has garnered considerable support among arms control specialists, but since the campaign was first launched the mainline churches have been by far the most important force promoting the proposal. By May 1983, when the House adopted its freeze resolution, the idea had been endorsed by the United Presbyterians, the United Methodist Church, the Episcopal Church, and the American Baptist Churches. The Unitarian Universalist Association, the Union of American Hebrew Congregations (which represents Reform Jews), and the 1,200-member Rabbinical Assembly (which speaks for Conservative Judaism) also had endorsed the proposal.

Many of the churches have adopted extensive statements on nuclear weapons, and of these the most influential has been the pastoral letter the U.S. Conference of Catholic Bishops adopted in May 1983.[16] In the letter, which was drafted by a committee headed by Cardinal Joseph Bernardin of Chicago, the bishops called nuclear war immoral; rejected first use of nuclear weapons (even though U.S. policy for the defense of Western Europe calls for first use if necessary); and endorsed the doctrine of nuclear deterrence only on a "strictly conditional" basis — conditional, that is, on the pursuit of serious arms reduction. The bishops repeatedly emphasized their conviction that nuclear deterrence is not acceptable as a permanent strategy and that nuclear weapons eventually must be totally abolished. This view of mutual assured destruction as a provisional arrangement is essentially the same as the position adopted by the German Lutherans in the late 1950s, and it is consistent with the Vatican's position.

Much of the ferment in the U.S. Catholic Church can be traced, indeed, to the Vatican Council or "Vatican II" (1962-1965), which was called by Pope John XXIII to "renew" the church. The council had a strong impact on a large proportion of the church's current bishops during their formative years. Between 1973 and 1980, when Archbishop Jean Jadot was the Vatican's apostolic delegate in the United States, more than half of the current bishops were appointed. During this period the hierarchy took on a much more liberal or progressive cast. Catholic resistance to the war in Vietnam contributed to change in the church, and the Berrigan brothers also played a part. As priests, Philip and Daniel Berrigan pioneered new forms of civil disobedience — pouring blood on draft files, for example — that captured the imaginations of people in all branches of the movement against the Vietnam War.[17]

Efforts to Discredit Movement

During the second year of the Catholic Church's deliberations about nuclear war, the Reagan administration made a concerted effort to influence the bishops. On Aug. 3, 1982, speaking at the centennial meeting of the Supreme Council of the Knights of Columbus in Hartford, Conn., Reagan asked Catholics to reject the idea of a nuclear weapons freeze. When the bishops met in November to consider the second draft of their letter, William P. Clark — Reagan's national security adviser at that time and a Catholic — submitted a letter taking issue with many points. Clark hailed the third draft as a substantial improvement, but the bishops resented the implication that they had made significant changes in response to political pressure, and when they met in Chicago in May 1983 they restored some of the tougher language that had been toned down.[18]

Reagan also is reported to have sent Vernon Walters, a highly experienced diplomatic trouble-shooter, to the Vatican to see whether the pope might rein in the American bishops on nuclear issues.[19] Walters had dealt with Latin American leaders for Reagan, but his Vatican mission seems to have been unsuccessful. Although the pontiff has many important disagreements with the American bishops, he is considered to be in sympathy with them on the nuclear issue. Reagan's decision in January 1984 to restore diplomatic recognition to the Vatican may have been connected, however, with a desire to keep lobbying the pope on nuclear politics.

If Reagan's attempts to influence Catholics may be said to have met with mixed success, the same goes for his efforts with Protestants. On March 8, 1983, Reagan made a strong appeal to the National Association of Evangelicals meeting in Orlando, Fla. After reminding the evangelical Christians of his support for their positions on birth control clinics, school prayer, and abortion, Reagan urged them to "speak out against those who would place the United States in a position of military and moral inferiority" against communist countries, which he referred to as "the focus of evil in the modern world." Reagan's audience gave him a standing ovation, but the speech nonetheless was less than a total success. The Association of Evangelicals, which represents 3.5 million members, including Mennonites and Brethren, declined to endorse Reagan's policies, sticking to a centrist position.[20]

In November and December 1982, Reagan suggested in press conferences that activists working for a freeze were dupes of the Soviet KGB. When he was pressed for evidence to support this charge, which the FBI dismissed in a report issued three months later, the White House referred reporters to some government studies and some magazine articles. On inspection, it seemed that the main source for Reagan's allegations had been an article that appeared in the October 1982 issue of *Reader's Digest.*[21]

The main American target of the *Reader's Digest* article was Provance, the AFSC's top peace organizer from 1974 to 1982, who cultivated exceptionally close and extensive contacts with Europeans. Ironically, Provance and other leaders of independent peace organizations in Europe and the United States say they were accused in earlier years of being CIA dupes by Soviet officials and officials connected with the international peace organization supported by the Soviet Union, the World Peace Council.

When it first appeared, the West European peace movement met with an ambivalent response from Soviet authorities. The Soviets plainly were eager to manipulate the movement to support their position in arms control negotiations, and there is evidence that they tried to channel money into favored organizations in the movement. But the independence of the movement, and especially the tendency of its leaders to condemn the arms policies of both superpowers in the same breath, made them nervous. Probably they feared, with good reason, that the West European movement might inspire similar activism in the East.

By 1981-1982 independent peace activism had in fact appeared in East Europe, most prevalently in the Democratic Republic of Germany. East German organizers operated under the wing of the Protestant church, which has a privileged status in the society, rather as the Catholic

Church does in Poland, and their main targets were military training and conscription. They wanted the government to give full recognition to conscientious objectors, and they wanted to end military training in public schools or at least enable students to opt out of such training. Similar efforts have appeared, albeit on a much smaller scale, in the other Eastern Bloc countries, including the Soviet Union.

Through 1983, when the Soviet Union and East European authorities were anxious to stay on reasonably good terms with West European activists, independent peace activities were tolerated in countries such as East Germany. But after the West German parliament voted to endorse the deployment of Pershing II and cruise missiles, which began in December 1983, the East German authorities reportedly cracked down hard on peace activists. In many cases activists had the choice of going to jail or leaving the country, and apparently many left.[22]

The deployment of new Soviet missiles in East Germany and Czechoslovakia, a move taken in retaliation against NATO's stationing of new missiles at the end of 1983, has met with criticism from citizens and officials in East Europe. For example, 24 Czech workers sent the country's leader, Gustav Husak, a petition opposing new Soviet missiles and demanding that existing missiles of a similar kind be dismantled. "It was the Soviet Union that began with the deployment of the missiles and with constant arming, although it had not been and is not threatened by anyone," the workers said, according to a report in *The New York Times* on Dec. 28, 1983. In Saxony, a state in East Germany, the synod of the Evangelical Church adopted a statement saying "we hold fast and say 'no' to new missiles." East German officials have expressed concern about the cost of the new missiles, which the Soviet Union expects its allies to help cover. The leaders of Bulgaria and Romania, who often take independent foreign policy positions, have criticized the new Soviet deployments.

Militarism: East and West

In the view of E. P. Thompson, an eminent British historian who is one of the dominant intellectual figures connected with CND and a spokesman for the closely affiliated campaign for European Nuclear Disarmament (END), militaristic tendencies in both the Soviet Union and the United States have fed on one another since 1945, keeping the Western world in a permanent cold war. Serious disarmament measures cannot be expected from the governments that have the strongest vested interest in the arms race, Thompson believes, and he is convinced that the two systems are united "in mutual hostility to any genuine non-alignment, 'neutralism' or 'third way.'" END's central objective, in Thompson's opinion, is to begin putting Europe's common culture "back into one piece" by initiating a process that would lead toward the "dissolution of both blocs."[23]

People who have worked to promote bilateral and multilateral arms control take a highly skeptical view of Thompson's notion that more is to be gained from unilateral disarmament measures forced upon governments in the West and East than by protest movements operating at the grass roots. Jeremy J. Stone, staff director of the Federation of American Scientists, has described the idea that the peace movement in the West might be matched by a similar movement in the East as naive and perhaps even dishonest.[24] To many Americans, it seems unfair to talk about U.S. and Soviet militarism in the same breath, as though U.S. society were as regimented as the Soviet Union's. After all, they point out, protests against military policies are unknown and illegal in the Soviet Union, while in the United States independent action by citizens has influenced decisions in many cases. In addition to the mass movement against the Vietnam War, popular opposition helped minimize civil defense efforts in the early 1960s, defeat the anti-ballistic missile (ABM) in the late 1960s, and delay the B-1 and MX in the 1970s and early 1980s.

Interestingly, the Russians — including dissidents and exiles — tend to think that Thompson's view of the superpowers is unfair to the Soviet Union. In a rejoinder to Thompson published in *The Nation* in 1982, Zhores and Roy Medvedev maintained that the military-industrial complex is "under much less responsible control in the United States than in the USSR."[25] The Medvedev brothers argue that the party's supremacy over the military is absolute in the Soviet Union, while in the United States defense contractors and Pentagon officials exert an excessive influence over policy.

Independent authorities on military trends generally agree that an urge for profits and employment plays a bigger part in U.S. military policy than it does in the Soviet Union. The difference shows up in the intense lobbying seen every year over the U.S. defense budget, which fluctuates frequently, and in arms exports, which capitalist countries often permit or even promote for strictly commercial reasons. But the difference appears to be mainly a matter of degree. The Soviet Union currently exports more arms than the United States does, and experts are beginning to detect economic motives in Soviet decisions to sell weapons.[26] As for the party's control of the military, the unquestioned supremacy of the civilian leadership may owe a good deal to respect, privileges, and power it accords the army, air, and naval forces. The military leadership has played a major role in every succession struggle since the death of Stalin, and whenever the civilian leadership is closely divided, the military establishment stands a good chance of exerting decisive influence.

People in the Soviet Union are reminded almost daily of how the Red Army saved Russia from the barbaric German invaders in what is officially called the "Great Patriotic War." Political-military training and indoctrination is extensive in the schools, and officers are treated with virtually universal respect. They are among the more privileged members of society, and they have an almost complete monopoly on discussion of the technical issues that affect their interests. In contrast to the United States and other Western nations, there is no academic community of defense experts who makes it its specialty to advise and criticize politicians on technical military questions. This probably is one reason, together with the Russian phobia about foreign invasion, why the Soviet government has put so much emphasis on ABM systems and civil defense, despite their poor prospects for success. It also helps account for the military's reluctance to completely accept the notion that nuclear war would be absolutely unwinnable.[27]

According to David Holloway, the author of a 1982 book on Soviet defense policy, "there is nothing to suggest that the Soviet leaders think that a general nuclear war would be anything other than catastrophic, for the victors as well as the vanquished."[28] Holloway sees no evidence that the Soviet Union is an intrinsically expansionist country, although he agrees with other scholars that the Soviet political elite expects its influence in world affairs to be recognized commensurate with the country's status as a great military power. That may help explain the Soviet Union's stubborn insistence on its right to persist as a "great power" in its costly and unpopular Afghanistan war. In Afghanistan, for the first time since World War II, large numbers of Russians have been killed or wounded in combat.

With the World War II generation on the verge of passing from the scene, much speculation among Sovietologists has centered on what their successors will be like. Generally, the Soviet Union's successor generations are expected to be more cosmopolitan and less xenophobic

227

than the current leaders.[29] They may, however, also be less cautious and more aggressive in their foreign policies, having not experienced the horrors of World War II firsthand. Andrei Sakharov, the Soviet Union's best known dissenter, wrote several years ago that he liked "this new layer of leaders coming to the top even less than its predecessors. . . . There is a dreadful cynicism, careerism, and a complete indifference to ideals in international affairs."[30]

When Leonid Brezhnev's health began to deteriorate in the late 1970s, the Soviet Union's military leaders began to assume more prominence in the government's public activities. When Brezhnev and Jimmy Carter met to sign the SALT II agreements, for example, Defense Minister D. F. Ustinov was very much in evidence. After Brezhnev died and the ailing Yuri Andropov took over power, military leaders became even more prominent, especially whenever arms control issues were involved. Civilians have remained firmly in control since Andropov's death, but ever since a military government took over in Poland at the end of 1981, experts on the Soviet Union have become more cautious about predicting the future of civil-military relations.

In the United States, as in the Soviet Union, outright military rule is almost unthinkable. At the same time the United States clearly is a much less militaristic country as far as the military's influence in society is concerned. The military is not considered an especially prestigious profession among many Americans, and there is a long tradition of looking upon standing armed forces as a regrettable necessity.

For all that, the U.S. government has been much readier than the Soviet Union's to use military force since World War II, as studies prepared by scholars at the Brookings Institution showed in the late 1970s.[31] Between 1945 and 1975 the United States mounted 215 limited military operations (not counting wars and arms exports), or 7.2 a year; the Soviet Union between 1944 and 1979 conducted 187 similar operations (not counting arms transfers), or 5.3 a year. Moreover, about three quarters of the Soviet actions were taken to consolidate and preserve its position in East Europe, while three quarters of the U.S. actions were in the Third World.

Since 1945 political pressure repeatedly has been for major buildups of the American arsenal, and in both the late 1950s and the late 1970s a leading rationale for higher U.S. defense spending was to contain Soviet influence in the Third World. While attitudes toward defense spending vary widely among Americans, every time a military buildup has occurred

Regional Differences in 1981 Vote on
Proposed 2 Percent Cut in Defense Budget

Vote by Region	House			Senate	
	Yeas	Nays	Percent	Yeas	Nays
Total	197	202	49%-51%	36	57
East	68	39	64 -36	12	11
South	19	92	17 -83	5	19
Midwest	76	39	66 -34	15	9
West	34	32	52 -48	4	18

Vote by Sub-Region and Party

	Yeas	Nays	Percent	Yeas	Nays
New England	18	3	86%-14%	7	5
Democrats	11	1	92 - 8	4	2
Republicans	7	2	78 -22	3	3
Mid-Atlantic	50	36	58 -42	5	6
Democrats	33	14	70 -30	2	4
Republicans	17	22	44 -56	3	1
South	19	92	17 -83	5	19
Democrats	15	57	21 -79	2	13
Republicans	4	35	10 -90	3	6
Industrial Midwest	59	25	70 -30	6	4
Democrats	37	5	88 -12	3	3
Republicans	22	20	52 -48	3	1
Farm Midwest	17	14	55 -45	9	5
Democrats	7	5	58 -42	2	2
Republicans	10	9	53 -47	7	3
Mountain and Desert	7	11	39 -61	1	12
Democrats	5	2	71 -29	0	4
Republicans	2	9	18 -82	1	8
Pacific	27	21	56 -44	3	6
Democrats	24	4	86 -14	2	2
Republicans	3	17	15 -85	1	4

Source: Jenny Greene, Congressional Quarterly Research Department.

the pressure has come primarily from southern and southwestern Democrats and from western or southwestern Republicans. For example, Lyndon B. Johnson of Texas worked closely with Richard B. Russell of Georgia to promote higher defense spending after the Korean War broke out and again after Sputnik was launched. When Johnson was president, Russell was chairman of the Senate Armed Services Committee and Rep. Carl Vinson, D-Ga., presided over the House Armed Services Committee. Since Reagan has been president, Sen. John Tower, R-Texas, has been chairman of the Senate Armed Services Committee; if the Democrats were to regain control of the Senate, John Stennis, D-Miss., would take over the committee. Traditionally, southerners have had control of defense committees because of their seniority.

On close votes concerning defense spending, arms control, or military action, a coalition of southern Democrats and western Republicans almost always is ranged against a looser coalition of doves or pragmatists that typically include western Democrats and representatives of both parties from the Middle West and Northeast. This pattern of voting would seem to be loosely related to the structure of the U.S. procurement industries, but other factors also play a role. For reasons connected with their unique history within the United States, southerners tend to regard the military more highly than other Americans.[32] Middlewestern Republicans rely heavily on a constituency of small businessmen and farmers who have been suspicious of foreign involvements and hostile to the military life, which they regard as idle and unproductive. Western Republicans have been close to the big businesses that have been the most rapidly expanding and aggressive sector of the U.S. economy since 1945.

The Danger of Stereotypes

In any discussion of basic trends in countries as complex as the Soviet Union and the United States, there is a great danger in locking any particular group or institution into a stereotype. In recent discussions about arms control, Sen. Sam Nunn, D-Ga. — Carl Vinson's nephew, who currently holds Russell's seat — has promoted a proposal that was supported by Sen. Gary Hart, D-Colo., George McGovern's campaign manager in 1972 and a presidential candidate in 1984. In Nunn's "builddown" proposal, each superpower would retire two old nuclear weapons for every new one it builds. In the 1983 debate over the MX missile in the United States, supporters of the missile were surprised

1981-1982 Defense Outlays by State*

(in dollars, per capita)

	1981	1982	In-crease		1981	1982	In-crease
New England				Mich.	88	115	27
Conn.	143	238	95	Ohio	195	244	49
Maine	252	340	88	Wis.	48	66	18
Mass.	220	294	74				
N.H.	447	716	269	**Farm**			
R.I.	328	644	316	**Mid-West**			
Vt.	102	132	30				
				Iowa	82	113	31
Mid-Atlantic				Kan.	326	421	95
Del.	349	444	95	Minn.	73	108	35
Md.	560	789	229	Mo.	252	485	233
N.J.	90	129	39	Neb.	303	410	107
N.Y.	90	129	39	N.D.	463	605	142
Pa.	179	254	75	S.C.	278	385	107
W.Va.	54	76	22				
South				**Mountain**			
Ala.	465	560	95	**and Desert**			
Ark.	198	264	66				
Fla.	374	542	168	Ariz.	433	560	127
Ga.	520	677	157	Colo.	560	1,815	1,255
Ky.	311	367	56	Idaho	215	284	69
La.	217	281	64	Mont.	234	290	56
Miss.	337	463	126	Nev.	539	757	218
N.C.	365	460	95	N.M.	656	901	245
Okla.	650	802	152	Utah	752	950	198
S.C.	539	713	174	Wyo.	306	433	127
Tenn.	160	254	94				
Texas	423	545	122	**Pacific**			
Va.	1,255	1,696	441				
				Alaska	1,795	2,339	544
Industrial				Calif.	477	607	130
Mid-West				Hawaii	1,604	1,965	361
Ill.	186	254	68	Ore.	79	109	30
Ind.	115	152	37	Wash.	453	590	137

* Figures do not include procurement.

Source: Northeast Midwest Congressional Coalition

when the governors of the arch-conservative Mormon Church came out against its deployment. They said that its deployment in Utah and Nevada would be incompatible with the church's gospel of "peace to the peoples of the earth," and in a statement on the giant mobile missile they asked the country's leaders to "marshal the genius of the nation to find viable alternatives which will secure at an earlier date and with few hazards the protection from possible enemy aggression which is our common concern."[33]

A similar surprise occurred in December 1983 during a press conference in Moscow with Marshal Nikolai V. Ogarkov, chief of the Soviet Union's General Staff. When a reporter asked him if he had seen "The Day After," ABC's film about nuclear war, he said, "I have seen the film and I believe that the danger it depicts is real." With his voice reportedly assuming an urgent edge he continued:

> Judge for yourself. The strategic nuclear forces of the United States can fire at a single launching 12,000 warheads with a total yield of 3,400 megatons. This is 170,000 times more than the yield of the first atomic bomb that the United States dropped on Hiroshima. . . . Moreover, this is only part of the story. If we add to this total the retaliatory capacity of the Soviet Union, which can hardly have fewer nuclear systems at its disposal, I think the matter speaks for itself. We have reached the point when it is time to put an end to nuclear madness. The situation, as it is, not only makes no sense. It is very, very dangerous.

No doubt many Americans would dismiss General Ogarkov's statement as yet another Soviet attempt to favorably impress peace activists, and many Soviets would make the same claim about Senator Nunn's builddown proposal. However that may be, the peace movement is a force that leaders of the superpowers can no longer ignore.

Notes

1. See Harold A. Feiveson, "The Dilemma of Theater, Nuclear Weapons," *World Politics* (January 1981): 282-298.
2. See David Holloway, *The Soviet Union and the Arms Race* (New Haven: Yale University Press, 1982), 66.

3. See Stockholm International Peace Research Institute, *Tactical Nuclear Weapons* (London: Taylor & Francis, 1978), 237.

4. See, for example, General Sir John Hockett and other top-ranking NATO generals and advisers, *The Third World War: August 1985* (New York: Macmillan Publishing Co., 1978).

5. See Stockholm International Peace Research Institute, *Tactical Nuclear Weapons*, 26; also Paul F. Walker, "New Weapons and the Changing Nature of Warfare," *Arms Control Today* (April 1979).

6. Feiveson, "The Dilemna," 296-297.

7. See Daniel Cohn-Bendit and Gabriel Cohn-Bendit, *Obsolete Communism: The Left-Wing Alternative* (New York: McGraw-Hill, 1968).

8. William Sweet, "Europe's Postwar Generations" in *Editorial Research Reports, 1981, Vol. II* (Washington, D.C.: Congressional Quarterly, 1982), 933.

9. See Horst Mewes, "The West German Green Party," *New German Critique* (Winter 1983): 55-59.

10. For examples, see Robert Scheer, *With Enough Shovels: Reagan, Bush and Nuclear War* (New York: Random House, 1982).

11. For an interview with Szabo see William Sweet, "Europe's Postwar Generations," 939-940. Szabo has summarized his findings and views in "The Successor Generation in Europe," *Public Opinion* (February-March 1983), 9-11. The book he edited is *The Successor Generation in Europe: International Perspectives of Postwar Europe* (Woburn, Mass.: Butterworth Publishers, 1983). For other perspectives see Pierre Hassner, "The Shifting Foundation," *Foreign Policy:* 3-20, and the Atlantic Council, "The Successor Generation" (January 1981).

12. See "Die Weinerliche Generation," *Welt am Sontag,* Feb. 13-March 13, 1983. Two more sophisticated treatments of the same themes are to be found in Gerd Langguth, *Protestbewegung: Die Neue Linke Seit 1968* (Cologne: Verlag Wissenschaft und Politik, 1983), and *Jungend Ist Anders* (Freiburg: Herder, 1983). Langguth, a former Christian Democratic member of West Germany's parliament, currently is director of the government's Office for Political Education.

13. William Sweet, "Europe's Postwar Generations," 939.

14. See William Sweet, "The 1982 Election," *Bulletin of the Atomic Scientists* (January 1983): 56-57.

15. See Harold Feiveson and Frank von Hippel, "The Freeze and the Counterforce Race," *Physics Today* (January 1983): 36-49.

16. See National Conference of Catholic Bishops, Ad Hoc Committee on War and Peace, "The Challenge of Peace—God's Promise and Our Response," (Washington, D.C.: The Confraternity of Christian Doctrine, 1983); also,

Bruce L. van Hoorst, "The Churches and Nuclear Deterrence," *Foreign Affairs* (Spring 1983): 827-852.

17. See Francine du Pessix Gray, *Divine Disobedience* (New York: Random House, 1969), and Mel Piehl, *Breaking Bread* (Philadelphia: Temple University Press, 1982). Even Catholics who disapproved of their beliefs and tactics could take some pride that the Berrigans were keeping alive a radical Catholic tradition established in this country by Dorothy Day and the Catholic Worker Movement.

18. See the interview with the Rev. J. Bryan Hehir, *The Inter Dependent* (July-August 1983).

19. William Sweet, "Christian Peace Movement," in *Editorial Research Reports, 1983, Vol. I* (Washington, D.C.: Congressional Quarterly, 1983), 363.

20. See Beth Spring, "Reagan Courts Evangelical Clout Against Nuclear Freeze," *Christianity Today*, April 8, 1983.

21. See John Barron, "The KGB's Magical War for Peace," *Reader's Digest* (October 1982): 206-259.

22. See Suzanne Gordon, "The View from the East," *Nuclear Times* (January 1984): 24-25.

23. See E.P. Thompson, *Beyond the Cold War: A New Approach to the Arms Race and Nuclear Annihilation* (New York: Pantheon Books, 1982), 74, and "East-West—Is There a Third Way?" *The Nation*, July 10, 1982, 49.

24. See Jeremy J. Stone, "European Disarmament Movement Really Wants Unilateral Disarmament," *F.A.S. Public Interest Report* (December 1981).

25. Roy A. Medvedev and Zhores A. Medvedev, "Nuclear Samizdat," *The Nation*, Jan. 16, 1982.

26. See Raimo Väyrynen, "Economic and Political Consequences of Arms Transfers to the Third World," *Alternatives* (VI, 1980): 131-155.

27. See David Holloway, "War, Militarism and the Soviet State," *Alternatives* (VI, 1980), 59-92.

28. Holloway, *The Soviet Union and the Arms Race*, 43.

29. See Jerry F. Hough, *Soviet Leadership in Transition* (Washington, D.C.: The Brookings Institution, 1980) and Seweryn Bialer, *Stalin's Successors: Leadership, Stability and Change in the Soviet Union* (New York: Cambridge University Press, 1980).

30. Quoted in Robert Conquest, "Worse to Come?" *The New Republic*, Jan. 17, 1981, 31.

31. See Barry M. Blechman and Stephen S. Kaplan, *Force Without War: U.S. Armed Forces as a Political Instrument* (Washington, D.C.: The Brookings Institution, 1978) and Stephen S. Kaplan, *Diplomacy of Power: Soviet Armed Forces as a Political Instrument* (Washington, D.C.: The Brookings Institution, 1981).

32. See Charles O. Lerche, *The Uncertain South: Its Changing Pattern of Politics in Foreign Policy* (New York: Quadrangle Books, 1964).

33. William Sweet, "MX Missile Decision" in *Editorial Research Reports, Vol I, 1981* (Washington, D.C.: Congressional Quarterly, 1981), 414.

Defense Secretary Caspar Weinberger testifies with Joint Chiefs of Staff Chairman John W. Vessey before a Senate Armed Services Committee hearing in February 1983.

Chapter 12

CONCLUSION: EFFECTS OF THE NUCLEAR AGE

Disturbed by the collapse of arms control, the deterioration of U.S.-Soviet relations, and the growing dissension in West Europe, American specialists on defense policy began to grope in the early 1980s for new proposals to reduce tensions. In a widely noted article published by *Foreign Affairs* magazine in spring 1982, four U.S. experts suggested that the United States should renounce first use of nuclear weapons, which would be a radical change of policy.[1] Since the 1950s the U.S. plan has been to respond with nuclear weapons to any attack on West Europe by the Soviet Union and its Warsaw Pact allies. One of the experts who proposed adoption of a "no first use" policy was Robert S. McNamara, who was defense secretary in the Kennedy and Johnson adminstrations. In 1983 McNamara went even further and proposed that NATO should renounce use of nuclear weapons altogether.[2]

The standing U.S. plan to use nuclear weapons if necessary to repulse a conventional Warsaw Pact attack has given the Soviet Union a propaganda advantage that it no longer would be able to exploit if the United States adopted a "no first use" policy. But it is questionable whether a mere declaration would be perceived in Europe as a truly meaningful gesture. To give substance to such a declaration, it might be necessary to remove nuclear weapons from European soil and reduce naval deployments on the periphery of Europe. In 1982 an independent commission on security and disarmament proposed that a zone in central Europe should be cleared of battlefield weapons. The commission, which was chaired by Olof Palme, the prime minister of Sweden, also advocated strengthening the United Nations to make it more capable of dealing effectively with threats to peace.

Critics of denuclearization proposals often contend that it would be necessary to spend more on conventional arms if stocks of nuclear weapons were sharply reduced. If, indeed, atomic arsenals were elimi-

237

nated altogether, defenders of the status quo argue that the only possible effect would be to return us to the nasty and brutish pre-nuclear world, in which truly frenetic conventional arms races and devastating wars were regular occurrences. Seen from this point of view, nuclear weapons — precisely because of their unthinkable destructiveness — represent, on balance, a force for good.

The notion that the bomb basically is a peacemaker seems to have been first formulated right after World War II at a conference held at the University of Chicago to discuss the implications of nuclear power. In a paper prepared for the conference the economist Jacob Viner said that in the nuclear age "retaliation in equal terms is unavoidable and in this sense the atomic bomb is a war deterrent, a peace-making force." [3] Viner, a scholar widely respected for his sober lucidity, once wrote that "pacifism in general, like fertilizer, is good only when evenly spread." [4] Spread unevenly, he implied, pacifism is just manure.

Echoing Viner's viewpoint, French president Francois Mitterand said in 1983 that the trouble with the anti-nuclear movement was that "the missiles are in the East, but the demonstrators are in the West." Mitterand spoke for a broad body of opinion, which holds that people in the peace movement are unreasonable and "angst ridden." To an extent European anti-nuclear activists may have fed such criticism by frequently claiming that they alone "feel the fear." Their position is that it is irrational in today's nuclear-armed world not to be afraid; they think that the bomb should change our ways of feeling and not just our ways of thinking, as Einstein suggested.

Among those who regard themselves as experts of foreign affairs and other matters of high policy, it always has been tempting to dismiss critics as ill-informed, naive, confused, and overly emotional. At one time seemingly arcane matters such as foreign policy and energy policy were in fact very much the reserve of the country's intellectual, political, and business elites. Members of the "establishment" would meet regularly at private clubs and in exclusive organizations such as the Council on Foreign Relations to discuss the "national interest." Exceptionally enterprising citizens might write to their representatives in Congress about national issues, but otherwise there was little opportunity at the grass-roots level to develop and promote foreign policy positions.

Today the situation is drastically changed in that U.S. citizens need go only as far as their local church group or their mailbox to get involved in matters of high policy. By the early 1980s anybody who ever gave a dime to a public organization was likely to be receiving a veritable avalance of "direct mail" asking for support in campaigns connected with nuclear weapons and nuclear power. Often such letters conjured up grim visions and made grand claims about the difference a small contribution might make to the fate of the earth.

At the end of 1983, for example, the U.S. Committee Against Nuclear War — a political action committee sponsored by some members of Congress — wrote to selected citizens that as the "New Year arrives . . . we are forced to ask: Will this be the last one?" As little as $25, the letter said, would help make "this year THE YEAR of the Freeze, THE YEAR to stop MX, THE YEAR to dismantle the European missiles."

SANE, the "committee for a sane nuclear policy," sent out letters around the same time saying that "our nation seems locked in a relentless march toward war." Public Citizen, one of Ralph Nader's organizations, informed its mail recipients that "top government officials" had been keeping "a dirty little secret from the American people" — the secret that the atomic bomb has a "twin brother . . . atomic power." Ground Zero, an organization that has concentrated mainly on educating citizens about the horrors of nuclear war, warned that "time is running out," but that "we can stop the threat of nuclear holocaust," and that "Ground Zero is the most effective way to help you do just that."

Apocalyptic rhetoric and sweeping claims naturally arouse skepticism even among the people most friendly to the cause of nuclear disarmament. Writing in the March 1984 issue of *The Progressive* magazine, Samuel H. Day, Jr., warned readers that giving money to groups that are well-intentioned but ineffective might make people feel better, the way a placebo sometimes does, but that it would not do anything to solve real problems. "Before putting your money down on any program to save the world from nuclear extinction," Day counseled, it is sensible to ask "after the talking and meeting and voting are over, does it do anything?"

Almost everybody in the United States agrees that nuclear war would be a terrible catastrophe, and almost everybody agrees that the danger of nuclear war should be reduced as much as possible. But opinions differ drastically on how that should be accomplished, and it is of the utmost importance to be discriminating about the methods proposed. It also is important, most Americans would agree, to be wary about ascribing ulterior motives to people who take opposing positions.

239

At the same time it never hurts to be aware of how positions might be colored by self interest — and this goes not only for the constitutencies that stand to gain in some ways from a continuing arms race, but also for the groups that say they are working to end it.

During the Vietnam War a famous anti-war poster depicted a tree with a caption that read, "War is unhealthy for trees and other living things." An almost equally famous satire said, "War is healthy for poster makers and other living things."

Notes

1. McGeorge Bundy, George F. Kennan, Robert S. McNamara, and Gerard Smith, "Nuclear Weapons and the Atlantic Alliance," *Foreign Affairs* (Spring 1982): 753-768.
2. Robert S. McNamara, "The Mile Long Role of Nuclear Weapons," *Foreign Affairs* (Fall 1983): 59-80.
3. Quoted in Theodore Draper, "Nuclear Temptations," *The New York Review of Books,* Jan. 19, 1984, 41. For accounts of the Chicago conference see Fred Kaplan, *The Wizards of Armageddon* (New York: Simon & Schuster, 1983), 24-28, and David E. Lilienthal, *Journals: The Atomic Energy Years, 1945-1950* (New York: Harper & Row, 1964), 637-645.
4. See Jacob Viner, *New Perspectives on Peace*, Charles R. Walgreen Foundation Lecture (Chicago: University of Chicago Press, 1944).

SUGGESTED READINGS

The Nuclear Era

The most comprehensive and up-to-date account of the developments and decisions that led to the bombing of Hiroshima is Martin J. Sherwin's *A World Destroyed: The Atomic Bomb and the Grand Alliance* (New York: Alfred A. Knopf, 1975). Several of the older histories remain well worth reading, especially Robert Jungk, *Brighter Than a Thousand Suns* (New York: Harcourt, Brace & Co., 1958), and Herbert Feis, *The Atomic Bomb and the End of World War II* (Princeton, N.J.: Princeton University Press, 1970). For a widely noted attempt to visualize the effects of a nuclear war see Jonathan Schell, *The Fate of the Earth* (New York: Alfred A. Knopf, 1982). John Hersey's *Hiroshima* (New York: Alfred A. Knopf, 1946) is available in a Bantam paperback (New York: Bantam Books, 1979). For a general history of the atomic era see Ronald W. Clark, *The Greatest Power on Earth* (New York: Harper & Row, 1980), and Bertrand Goldschmidt, *The Atomic Complex: A Worldwide Political History of Nuclear Energy* (La Grange Park, Ill.: American Nuclear Society, 1982).

For critical or even muckraking descriptions of the nuclear industry and its development see Daniel Ford, *The Cult of the Atom* (New York: Simon & Schuster, 1982); Peter H. Metzger, *The Atomic Establishment* (New York: Simon & Schuster, 1972); Mark Hertsgaard, *Nuclear Inc.: The Men and Money Behind Nuclear Energy* (New York: Pantheon Books, 1983); Peter Pringle and James Spigelman, *The Nuclear Barons* (New York: Holt, Rinehart & Winston, 1982); and Irvin C. Bupp and Jean Claude Derian, *Light Water: How the Nuclear Dream Dissolved* (New York: Basic Books, 1978). For a survey of positions on nuclear energy see Michio Kaku and Jennifer Trainer, eds., *Nuclear Power: Both Sides* (New York: W.W. Norton & Co., 1982). Informative chapters on the role scientists have played in the nuclear debate are to be found in Joel Primack and Frank von Hippel, *Advice and Dissent: Scientists in the Political Arena* (New York: Meridian, 1976). The best introduction to the subject of nuclear terrorism is John McPhee's *At the Curve of Binding Energy* (New York: Ballantine Books, 1974), which is about Theodore B. Taylor, the world's foremost expert on the subject.

Nuclear Energy

One of the best books describing nuclear technology is Walter Patterson, *Nuclear Power* (New York: Penguin Books, 1983). For opposing views on the dangers of radiation see John W. Gofman, *Radiation and Human Health* (San Francisco: Sierra Club Books, 1981), and Ralph Lapp and George Russ, *Radiation Risks for Nuclear Workers* (Bethesda, Md.: Atomic Industrial Forum, 1979). On nuclear accidents see Daniel Ford and Steven J. Nadis, *Nuclear Power: The Aftermath of Three Mile Island* (Cambridge, Mass.: Union of Concerned Scientists, 1982). For a highly critical overview of nuclear power issues, which also discusses some aspects of nuclear proliferation, see Anna Gyorgy and others, *No Nukes: Everyone's Guide to Nuclear Power* (Boston: South End Press, 1979).

Nuclear Proliferation and the Arms Race

To a surprising extent, writing on nuclear proliferation has been the reserve of specialists. I am aware of no books that attempt to provide an overview of the problem for the general public. For a lively and absorbing treatment of the problem in one area, however, see Steve Weissman and Gerbert Krosney, *The Islamic Bomb: The Nuclear Threat to Israel and the Middle East* (New York: Times Books, 1982). The most readable histories of the strategic arms talks are John Newhouse, *Cold Dawn: The Story of SALT* (New York: Holt, Rinehart & Winston, 1973), and T. Strobe Talbot, *Endgame: The Inside Story of SALT II* (New York: Harper & Row, 1979). For a highly critical view of arms control see Alva Myrdal, *The Game of Disarmament* (New York: Pantheon Books, 1976). For a range of views on the arms race, U.S. defense policy, the military balance, and deterrence see John M. Collins, *Imbalance of Power: Shifting U.S.-Soviet Military Strengths* (San Rafael, Calif.: Presidio Press, 1978); The Boston Study Group, *The Price of Defense* (New York: Times Books, 1979); and The Harvard Nuclear Study Group, *Living With Nuclear Weapons* (New York: Bantam Books, 1982). For an argument on the benefits of nuclear deterrence see Bernard Brodie, ed., *The Absolute Weapon* (New York: Harcourt, Brace & Co., 1946), and Bernard Brodie, *War and Politics* (New York: Macmillan, 1973).

For discussions of the superpowers written from somewhat different perspectives see Richard J. Barnet, *The Giants* (New York: Simon & Schuster, 1977), and Adam B. Ulam, *The Rivals* (New York: Penguin Books, 1980). There is an enormous amount of critical literature on the U.S. "military-industrial complex," but by far the best book is Richard J. Barnet's *Roots of War* (New York: Penguin Books, 1973). For a recent treatment of the subject that stresses the interplay of Congress, the Pentagon, and defense contractors see Gordon Adams, *The Iron Triangle* (New York: Council on Economic Priorities, 1981). The origins of the current U.S. military buildup are explained from an antagonistic perspective in Jerry Sanders, *Peddlers of Crisis* (Boston: South End Press, 1983).

INDEX